XVIII International Seminars on Overarching Issues of the European Area

Volume Editors

Helena Pina
Felisbela Martins
Diogo Miguel Pinto
André Samora-Arvela

MDPI

Basel • Beijing • Wuhan • Barcelona • Belgrade • Novi Sad • Cluj • Manchester

Volume Editors
Helena Pina
University of Porto
Porto
Portugal

Felisbela Martins
University of Porto
Porto
Portugal

Diogo Miguel Pinto
University of Porto
Porto
Portugal

André Samora-Arvela
University Institute of Lisbon
Lisbon
Portugal

Editorial Office
MDPI AG
Grosspeteranlage 5
4052 Basel, Switzerland

This is a reprint of the Proceedings, published open access by the journal *Proceedings* (ISSN 2504-3900), freely accessible at: https://www.mdpi.com/2504-3900/113/1.

For citation purposes, cite each article independently as indicated on the article page online and as indicated below:

Lastname, A.A.; Lastname, B.B. Article Title. *Journal Name* **Year**, *Volume Number*, Page Range.

ISBN 978-3-7258-6091-3 (Hbk)
ISBN 978-3-7258-6092-0 (PDF)
https://doi.org/10.3390/books978-3-7258-6092-0

Cover image courtesy of Diogo Miguel Pinto

Contents

About the Editors

Helena Pina

Helena Pina completed her degree in Geography at the University of Porto in 1982, followed by a master's in human Geography at the University of Coimbra. She earned her PhD in Human Geography in 2004 from FLUP, with Distinction and Praise. In this context, she began her teaching career in 1982 as a Trainee Assistant at FLUP, transitioning to Assistant after completing her master's degree. After completing her PhD in Human Geography with the thesis titled "The Alto Douro, a contrasting space in mutation," she began her role as an Assistant Professor in 2004, achieving a permanent appointment in 2009. Throughout her academic career, she has shown a high commitment to teaching and pedagogical innovation, focusing mainly on social and economic geography, with special emphasis on the sustainable development of rural areas in Portugal. She has recently been researching resilient landscapes and forest fires. In an innovative commitment, she has taught, mostly as the course leader, subjects such as Rural Geography, Rural Dynamics in the European Union, European Space Issues, and various Technical Seminars. She is a researcher at the Centre for Geographic Studies and Spatial Planning (CEGOT); coordinating and/or participating as a researcher in several scientific projects, such as AVODIS as C-IR, PCIF/AGT/0054/2017; and currently coordinating the MONTTUR Project FLUP/PS-CPS/2024-81843, among other relevant projects. She is a member of several Scientific Associations, as well as the Scientific and Editorial Committee of several national and international reference magazines and publications. Her experience also includes organising congresses and seminars and supervising (or co-supervising) master's dissertations and PhD theses. This involvement and diversity of initiatives, both academic and professional, reflect her ability to lead and collaborate in multidisciplinary teams, addressing complex sustainable development issues.

Felisbela Martins

Felisbela Martins holds a PhD in Education Sciences (Faculty of Psychology and Educational Sciences University of Porto (FPCEUP) (2011)), a Master's Degree in Educational Sciences—Education, Development and Social Changes (FPCEUP) (1995), and a Bachelor's Degree in Geography (FLUP) (1979). Felisbela worked as a Guest Assistant Professor at the Department of Geography Faculty of Arts and Humanities of the University of Porto (1998-2015) and was a Member of the Secondary Education Department of the Ministry of Education and a Primary and Secondary Teacher of Geography. She retired in March 2022.

Diogo Miguel Pinto

Diogo Miguel Pinto holds a degree in Geography (2016) and a master's in Risks, Cities and Spatial Planning (2020) from the University of Porto, where he also completed postgraduate studies in Geographic Information Systems (2021). He later obtained a postgraduate diploma in Civil Protection Law (2023) from the University of Lisbon. He completed the Specialisation Course in Security, Defence and Climate Change (CESDAC 2025) and the XVI Civil Crisis Management Course at the National Defence Institute (IDN). His academic background bridges geography, risk management, and climate adaptation. He has worked as a geography technician and researcher at several institutions. He is a collaborating researcher at the Centre for Studies in Geography and Spatial Planning (CEGOT). His research focuses on rural innovation, sustainable development, landscape resilience, wildfire risk, and civil protection. He has coordinated international initiatives such as the UNESCO World Heritage Volunteers "Guardians of the Douro". He has also served as an external evaluator for A3ES, a certified trainer (IEFP), and a member of several scientific associations. He has authored and reviewed numerous scientific papers, book chapters, and conference proceedings, and has participated in over 200 academic events.

André Samora-Arvela

André Samora-Arvela holds bachelor's and master's degrees in Landscape Architecture from the Faculty of Science and Technology of the University of Algarve (UAlg); a postgraduate degree in Territorial Management, specializing in Remote Sensing and Geographic Information Systems (GIS); and a PhD in Climate Change and Sustainable Development Policies, specializing in Geography and Planning, from the Faculty of Social and Human Sciences of the New University of Lisbon (FCSH, UNL). André Samora-Arvela is currently an integrated member of the Information Sciences, Technologies and Architecture Research Center (ISTAR) of the University Institute of Lisbon (ISCTE -IUL) and member of the project "GreenCity4Aging: the effects of urban green streets on mobility, social integration and ageism among older people" (Ref.[a] 2022.03478.PTDC). He was a professor at the Faculty of Arts of the University of Porto and an integrated member of the Center for Geography and Spatial Planning Studies (CEGOT) from 2021 to 2023 as part of his post-doctorate; a collaborating member of the Interdisciplinary Center for Social Sciences (CICS.NOVA), Faculty of Social and Human Sciences (FCSH, UNL), from 2015 to 2021; and a collaborating member of the Center for Research on Space and Organizations (CIEO), which was renamed the Research Center for Tourism, Sustainability and Wellbeing (CinTurs), UAlg, from 2011 to 2021. His transdisciplinary vocation and eclecticism are reflected in his participation in international and national projects within many fields, including spatial planning; geography; landscape architecture; societal challenges of mitigation and adaptation to climate change; community preparedness for risk; integrated management of rural fires; space, active aging and ageism; sustainable development and globalization; sustainable tourism development; tourist experience; geographic information systems; and methods applied to social research.

proceedings

Editorial

Preface: XVIII International Seminars on Overarching Issues of the European Area: "Building Sustainable and Resilient Territories" †

Helena Pina [1,*], Felisbela Martins [1,*], Diogo Miguel Pinto [1,*] and André Samora-Arvela [2,*]

1 CEGOT, Faculty of Arts and Humanities, University of Porto, 4150-564 Porto, Portugal
2 ISTAR, University Institute of Lisbon (ISCTE-IUL), Av. Forças Armadas, 1649-026 Lisbon, Portugal
* Correspondence: mpina@letras.up.pt (H.P.); felisbela.martins@gmail.com (F.M.); dspinto@letras.up.pt (D.M.P.); anesamora@gmail.com (A.S.-A.)
† Presented at the XVIII International Seminars on Overarching Issues of the European Area, Porto, Portugal, 23–26 May 2024.

1. Conference Background

In the wake of the pandemic and amidst ongoing global conflicts, our interconnected world is grappling with significant challenges. Despite the immense threats and destruction inflicted on our planet, numerous ecosystems continue to exist, albeit suffering from critical biodiversity loss. It is common to observe dynamic, multifunctional territories thriving in some ecosystems, juxtaposed with others in decline, where the balance between development and sustainability is precarious. These regions bear the undeniable impacts of climate change, particularly exacerbated by human activities and the ongoing war in Europe.

Nevertheless, awareness of these pressing issues is growing. The environmental, economic, and social consequences of these challenges have emphasised the urgent need for action across multiple levels. As a result, strategies are being devised that prioritise inclusive development, focusing not only on conserving landscapes and cultural heritage but also on revitalising social structures. To achieve these objectives, the United Nations' 17 Sustainable Development Goals (SDGs), outlined in the 2030 Agenda, were defined as essential references. These goals represent critical imperatives in the current context, where warfare in the European region exacerbates social inequalities and environmental degradation.

2. Conference Theme and Topics

Emphasising SDG 11: Sustainable Cities and Communities, SDG 13: Climate Action, and SDG 15: Life on Land, this edition of *Proceedings* is a key output from the XVIII International Seminars on the Overarching Issues of the European Area, that took place from the 23rd to the 26th of May 2024 at the Faculty of Arts and Humanities, University of Porto, Portugal.

As part of the planet's ecosystem, humanity has inflicted significant harm through deforestation, habitat loss, and land degradation. Promoting sustainable use of ecosystems and protecting biodiversity is not merely an agenda; it is fundamental to our survival. Therefore, this Special Issue of *Proceedings* will centre on regional resilience, sustainable development, and land planning and management, with a focus on the connections between rural and urban areas in a changing climate.

This issue aimed to publish papers that explore key themes such as community resilience, strategic regional development scenarios, pathways for climate change adaptation, public policies for landscape planning, land-use change, and the essential factors of sustainable tourism. Contributions from scientists, experts, and practitioners in fields such as geography, risk and civil protection, land use and land cover, economics, rural and urban landscape planning, climate change, environmental and social challenges, tourism, and sustainable development constitute this Issue.

Many communications and posters were presented under the following topics: (1) Society and culture; (2) Education and teaching; (3) Social and economic dimensions of sustainability; (4) Population ageing and rural health; (5) Tradition, heritage, and tourism; (6) Economy and sustainability; (7) Circular economy and waste management; (8) Green technologies, energy, and innovation; (9) Entrepreneurship and digital transformations in territories; (10) Urban–rural connections; (11) Resilience and transitions in rural areas; (12) Sustainable agriculture and food systems; (13) Landscape and ecosystem management; (14) Biodiversity conservation and sustainability; (15) Resource management and conservation; (16) Climate change, environment, and sustainable development; (17) Vulnerabilities, risks and civil protection; (18) Environmental policy and governance; (19) Challenges of European policy and spatial planning; and (20) European security and defence.

3. Conference Organisers

Organising institution: Faculty of Arts and Humanities, University of Porto, Portugal.

Honour Committee:

- José Luís Carneiro—Minister of Internal Administration, XXIII Constitutional Government of Portugal
- Maria Leitão Marques—Member of the European Parliament, Brussels
- Frederico Francisco—Secretary of State for Infrastructure, XXIII Constitutional Government of Portugal
- Teresa Almeida—President, Commission for Coordination and Regional Development of Lisbon and Tejo Valley (CCDR-LVT)
- António Ceia da Silva—President, Commission for Coordination and Regional Development of Alentejo (CCDR Alentejo)
- Rui Moreira—Mayor of Porto
- Paulo Silva—Mayor of Mesão Frio
- Fátima Fernandes—Mayor of Montalegre
- José Manuel Gonçalves—Mayor of Peso da Régua
- Pedro Machado—Mayor of Lousada
- Miguel Nunes—Councillor of Lousada Municipal Council
- Isabel Ferreira Nunes—President, National Defence Institute (IDN), Portugal
- Maria Madalena dos Santos Alves—President, Foundation for Science and Technology (FCT), Portugal
- Tiago Oliveira—President, Agency for Integrated Management of Rural Fires (AGIF), Portugal
- Duarte Costa—President, National Authority for Emergency and Civil Protection (ANEPC), Portugal
- André Fernandes—National Commander for Emergency and Civil Protection (ANEPC), Portugal
- José Manuel Matias—Director Regional Adjunto, (DRAP-N), Portugal
- Pedro Moreira—President, Comboios de Portugal (CP), Portugal
- Rosário Alves—Executive Director, Forestis—Portuguese Forestry Association
- Luís Almeida—President, Vale do Douro Association
- Carlos Fonseca—Chief Technology Officer, FORESTWISE

- Duarte Caldeira—President, Centre for Studies and Intervention in Civil Protection (CEIPC)
- Paula Pinto—Director, Faculty of Arts, University of Porto
- Domingos Xavier Viegas—University of Coimbra
- Mary Cawley—National University of Ireland, Galway, Ireland
- Atanas Dermendzhiev—University of Veliko Tarnovo, Bulgaria
- Stella Dermendzhieva—University of Veliko Tarnovo, Bulgaria
- Ana Maria Bicalho—Federal University of Rio de Janeiro, Brazil
- Scott William Hoefle—Federal University of Rio de Janeiro, Brazil
- Ileana Constantinescu—Academy of Economic Studies, Bucharest, Romania
- Jerzy Makowsky—University of Warsaw, Poland
- Laura Sakaja—University of Zagreb, Croatia
- Borna Fuerst-Bjelis—University of Zagreb, Croatia
- Marta Arbinolo—OECD, France
- Fernanda Cravidão—University of Coimbra
- António Covas—University of Algarve
- Mercês Covas—University of Algarve
- Artur Cristóvão—University of Trás-os-Montes and Alto Douro
- Lúcio Cunha—University of Coimbra
- Luciano Lourenço—University of Coimbra
- Nicole Vareta—Centre of Studies in Geography and Spatial Planning, Portugal
- Hélder Marques—University of Porto

Scientific Committee:

- Adélia Nunes, University of Coimbra, Portugal
- Aleksandar Lukić, University of Zagreb, Croatia
- Ana Isabel Boura, University of Porto, Portugal
- Ana Paula Brandão, University of Minho, Portugal
- André Santos, Univ. Federal Rural Rio Janeiro, Brazil
- Andrews José de Lucena, Univ. Federal Rural do Rio de Janeiro, Brazil
- Annett Steinführer, Thünen Institute of Rural Studies, Germany
- António de Barros Cardoso, University of Porto
- Artur Costa, University Lusófona, Portugal
- Bruno Martins, University of Coimbra, Portugal
- Borna Fuerst-Bjelis, University of Zagreb, Croatia
- Brigite Botequim, Colab ForestWise, Portugal
- Célia Taborda, University Lusófona, Portugal
- Christine Fürst, Martin Luther University Halle-Wittenberg, Germany
- Cornelia Steinhäuser, Institute of Landscape Ecology, Germany
- Emmanuel Reynard, University of Lausanne, Switzerland
- Eric de Noronha Vaz, Metropolitan University, Canada
- Fátima Matos, University of Porto, Portugal
- Fátima Velez de Castro, University of Coimbra, Portugal
- Felisbela Martins, University of Porto, Portugal
- Francisco Antonio Navarro Valverde, University of Granada, Spain
- Francisco Proença Garcia, Portuguese Catholic University, Portugal
- Glória Teixeira, University of Porto, Portugal
- Hélène Roth, University of Clermont, France
- Hermínia Gonçalves, University of Trás-os-Montes and Alto Douro, Portugal
- Hossein Azadi, Gembloux Agro-Bio Tech, Universidade de Liège, Belgium
- Isabel Borges, Instituto Europeu de Estudos Superiores, Portugal

- Jesús Rodrigo Comino, University of Granada, Spain
- João Luís Fernandes, University of Coimbra, Portugal
- Joaquim Esteves da Silva, University of Porto, Portugal
- Joaquim Mamede Alonso, The Polytechnic Institute of Viana do Castelo, Portugal
- Joaquim Sande Silva, Agrarian School of Coimbra, Portugal
- Jorge Queiroz, University of Porto, Portugal
- Jorge Ribeiro, University of Porto, Portugal
- Jorge Ricardo Pinto, Higher Institute of Business and Tourism Sciences Portugal
- José António de Oliveira, The Polytechnic Institute of Porto, Portugal
- José Aranha, University of Trás-os-Montes and Alto Douro, Portugal
- José Azevedo, University of Porto, Portugal
- José Domingo Sánchez Martínez, Universidad de Jaén, Spain
- José Luís Zêzere, University of Lisbon, Portugal
- Karine Vargas, Universidade Federal Rural do Rio de Janeiro, Brazil
- Laura Sakaja, University of Zagreb, Croatia
- Luís Moreno, University of Lisbon, Portugal
- Luis Ramos, University of Trás-os-Montes and Alto Douro, Portugal
- Marcin Wójcik, University of Lodz, Poland
- Madalena Fonseca, University of Porto, Portugal
- Maria Elisabete Soares, Polytechnic Institute of Guarda, Portugal
- Maria Francisca Saraiva, University of Lisbon, Portugal
- Michael Woods, Aberystwyth University, United Kingdom
- Michele Salis, CNR, Institute of BioEconomy IBE, Italy
- Milen Penerliev, University of Shumen, Bulgaria
- Natália Ramos, Portuguese Open University, Portugal
- Nicole Devy Vareta, University of Porto, Portugal
- Norberto Santos, University of Coimbra, Portugal
- Paulina Tobiasz-Lis, University of Lodz, Poland
- Paulo Jorge Santos, University of Porto, Portugal
- Regina Cohen Barros, Univ. Federal Rural Rio Janeiro, Brazil
- Ronan Foley, Maynooth University, Ireland
- Sandra Lopes Luís, University of Lisbon, Portugal
- Sandra Oliveira, University of Lisbon, Portugal
- Slavi Dimitrov, Univ. Verliko Tarnovo, Bulgaria
- Suk Kyeong Kang, Niğde Ömer Halisdemir University, Türkie
- Ulrike Grabski-Kieron, University of Münster, Germany
- Valentina Bacciu, Institute of BioEconomy IBE, Italy
- Virgílio Borges Pereira, University of Porto, Portugal
- Virgílio Falco, University of Trás-os-Montes and Alto Douro, Portugal

Organising Committee:

- Ana Isabel Boura (University of Porto), Portugal
- André Samora-Arvela (University Institute of Lisbon), Portugal
- António de Barros Cardoso (University of Porto), Portugal
- Conceição Ramos (University of Porto), Portugal
- Diogo M. Pinto (University of Porto), Portugal
- Felisbela Martins (University of Porto), Portugal
- Helena Pina (University of Porto)—Chair, Portugal
- João Pedro Barreiros (University of Porto), Portugal
- José Luís Braga (European Institute of Higher Education), Portugal
- Jorge Ribeiro (University of Porto), Portugal

- Leandro Dias Oliveira (Federal Rural University of Rio de Janeiro), Brazil
- Lívia Madureira (University of Trás-os-Montes and Alto Douro), Portugal
- Maria José Roxo (NOVA University Lisbon), Portugal
- Paula Remoaldo (University of Minho), Portugal

Secretariats:

- Jéssica Fabiana Santiago Valente
- Filipa Dinis Mota
- Inês Dinis Mota
- Gonçalo José Cruz Ramos
- Paola Vicentini Boni
- Beatriz da Silva Martins
- Ana Carolina Ramalho Sampaio
- Maria Luísa Scharlau da Silva
- Célia Figueiras

Conflicts of Interest: The authors declare no conflicts of interest.

proceedings

Editorial

Statement of Peer Review †

Helena Pina [1,*], Felisbela Martins [1,*], Diogo Miguel Pinto [1,*] and André Samora-Arvela [2,*]

1 CEGOT, Faculty of Arts and Humanities, University of Porto, 4150-564 Porto, Portugal
2 ISTAR, University Institute of Lisbon (ISCTE-IUL), Av. Forças Armadas, 1649-026 Lisbon, Portugal
* Correspondence: mpina@letras.up.pt (H.P.); felisbela.martins@gmail.com (F.M.); dspinto@letras.up.pt (D.M.P.); anesamora@gmail.com (A.S.-A.)

† All proceeding papers published in this volume were presented at the XVIII International Seminars on Overarching Issues of the European Area, Porto, Portugal, 23–26 May 2024.

In submitting conference proceedings to *Proceedings*, the volume editors of the proceedings certify to the publisher that all papers published in this volume have been subjected to peer review administered by the volume editors. Reviews were conducted by expert referees to the professional and scientific standards expected of a proceedings journal.

- Type of peer review: double-blind;
- Conference submission management system: Conference Email;
- Number of submissions sent for review: 23;
- Number of submissions accepted for publication: 21;
- Acceptance rate: 91.3%;
- Average number of reviews per paper: 0.5;
- Total number of reviewers involved: 12;
- Any additional information on the review process: Our approach to manuscript review follows a double-masked process, ensuring that neither authors nor reviewers know each other's identities, thus minimising bias and promoting impartial evaluation.

Each paper submitted was evaluated by experts in the relevant field, based on the following criteria:

Originality: Contribution to the existing body of knowledge, novelty of data, and innovative approaches.
Relevance to conference themes: Alignment with the topics and objectives of the event and proceedings.
Scientific and methodological rigour: Soundness of the research design, validity of the methods, appropriateness of the analyses, and reproducibility.
Clarity and coherence of presentation: Logical flow, quality of writing, adequacy of figures/tables, and whether results support conclusions.
Contribution to the field: Potential impact on theoretical, empirical, or applied research.
Quality of references: Proper citation and use of up-to-date and relevant literature.

Following the review process, decisions included acceptance, minor, or major revisions, with final acceptance contingent on satisfactory response to reviewers' comments. All revised papers were re-evaluated before publication.

Proceedings **2025**, *113*, 22 https://doi.org/10.3390/proceedings2025113022

Acknowledgments: The authors would like to express their gratitude to all the reviewers for their invaluable support in reviewing the conference papers and the authors.

Conflicts of Interest: The authors declare no conflicts of interest.

proceedings

MDPI

Proceeding Paper

Rethinking Food and Farming in Europe †

Jean-François Hulot

Independent Researcher, 24100 Bergerac, France; jfhulot@me.com

† Presented at the XVIII International Seminars on Overarching Issues of the European Area, Porto, Portugal, 23–26 May 2024.

Abstract: Food security in Europe remains assured, but at too high an environmental cost. The old agro-chemical model has proven inadequate to meet the multi-faceted challenges of the 21st century. The most promising model must be sustainable and simultaneously improve economic performance, environmental protection, and social impact. Agro-ecology meets those goals, but its implementation would require a complete rethinking of EU policies for agriculture and food. The Common Agricultural Policy must radically change from quantitative to qualitative support and favour a demand-led rather than a supply-led approach to reflect consumer needs, not only in food pricing terms but also in nutritional and health aspects.

Keywords: common agricultural policy; farm to fork; agro-ecology; food systems; climate change; biodiversity loss; food security; human health

1. Introduction

Agro-ecology requires a complete rethinking of agriculture and food systems in the EU. The European Commission has already attempted to extricate the Common Agricultural Policy (CAP) from its corporatist feature via the European Green Deal, by integrating the entire food chain under a single strategy known as the "Farm to Fork" strategy [1]. It fixed targets such as a 50% reduction in pesticides use, which became controversial and was eventually opposed by the farming sector. Hence, the approach proved ineffective even before concrete implementation took place. Yet climate and biodiversity challenges remain more acute than ever. The CAP should be profoundly revised to help make food and farming more sustainable.

2. The Food and Farming Issue Is Inextricably Linked to Geopolitics, Environment and Climate Change

In 2005, the United Nations launched the concept of food security in its sustainable development goals (SDGs), more specifically in SDG 2 "Zero hunger", with the aim of ending hunger and food insecurity in the world by 2030. Food security for a country's population is not only a precondition for its development but also a geostrategic asset, strongly linked to preserving a country's strategic independence.

Against this background, the EU occupies the position of the world's leading agri-food exporter, as well as importer. In 2022, the EU exported EUR 229.1 billion worth of agricultural products and imported EUR 195.6 billion worth, generating a surplus of EUR 33.5 billion, a continuation of a robust and positive trend observed over the last 20 years as observed in Figure 1 [2]. Food security is largely assured in Europe, in line with the objectives assigned in 1962 for the CAP, which aimed at guaranteeing food supplies for the European population. Yields steadily increased thanks to seed selection, investment in machinery and intensification of farming by making wide use of chemical fertilizers and plant protection pesticides. Although this agro-chemical model has become the norm, it has not been without damaging consequences for water, air, soil and biodiversity through the pollution it causes (for example, contamination of water by nitrates, the residues of nitrogen fertilisers) and the destruction it generates (such as the collapse of pollinating insects due to the use of pesticides).

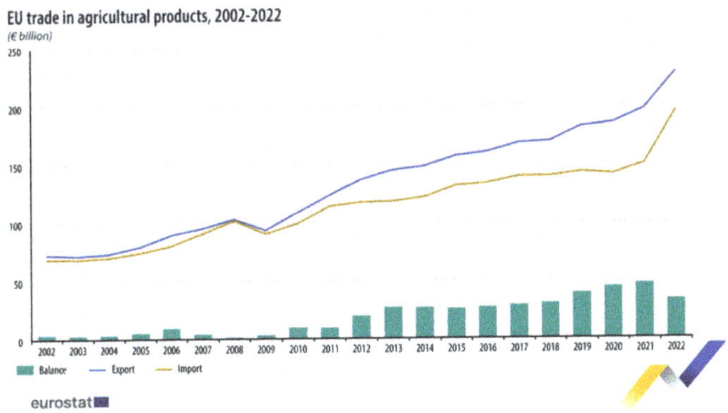

Figure 1. EU trade in agricultural products 2002–2022.

The European Environment Agency has identified agriculture as the main factor in the degradation of biodiversity and the destruction of natural capital in Europe, despite efforts to 'green' the CAP over the last 10 years [3]. Agriculture also has an impact on climate, mainly due to methane emissions from ruminants, and nitrous oxide emissions from animal waste and nitrogen fertilisers. The greenhouse gas (GHG) emitted from the agricultural sector had fallen thanks to efficiency gains, but is now relatively constant, dropping by only 2% since 2005. The EU agri-food system contributes 31% of EU total emissions.

3. The Real Economic and Strategic Priorities for Agriculture Have Dramatically Changed over the Past 30 Years

The CAP has provided support to farmers since the 1960s, with the initial objectives of boosting agricultural productivity, supporting an economically sustainable food supply, safeguarding fair prices and ensuring reasonably good farm income in the EU. Over time, priorities have evolved, placing increasing emphasis on supporting the provision of environmental public well-being, addressing externalities and promoting rural development. Despite the successive reforms the CAP underwent over the past 30 years in order to align it to evolving objectives, it has not succeeded in reducing GHG emissions neither restoring nor protecting its natural capital (soil, water, biodiversity).

While the CAP budget is still the EU's biggest item of expenditure with about EUR 55 billion per year, farm incomes remain lower than in other sectors. Three quarters of that budget is spent in the form of aid per hectare paid directly to farmers, forming the so-called "1st pillar of the CAP". This plays a vital role in determining farm income: direct aid varies according to production type and to the size of the farm (the more hectares a farm has, the higher the overall payment it receives from the EU budget) and can account for up to 100% of farm income in certain regions. The original purpose of this direct aid was to compensate for the fall in prices triggered by the 1992 CAP reform. In practice, this maintains the unequal situation between EU Member States, leading to complex correction mechanisms, particularly in favour of the Baltic States, and between large and small farms

(20% of farms in the EU receive 80% of aid [4]. They also encourage the concentration of farms and the continuation of unchanged agricultural practices.

The existence of substantial European subsidies is pushing up the price of agricultural land and driving down the prices paid to farmers by cooperatives and by the processing and distribution industries. Conversely, consumer prices are tending to rise, a sign that the sector's added value is in fact being captured by its downstream agri-food processors and supermarkets. Some countries, such as France and Spain, are trying through national legislation to pass on to primary producers a fairer share of the prices paid by consumers, but without much success ("Egalim" laws I, II, III... and "Ley de la cadena").

Today, agricultural production and environmental protection are often placed in opposition to each other, even though any form of agriculture designed for the long term needs healthy, functional ecosystems that respect natural cycles, inter alia water, nitrogen, phosphorus and carbon. The challenge now is to reconcile agriculture and the environment, both in practice and in people's minds, as agriculture must do more to protect the environment while adapting to a rapidly warming climate.

4. In the Future the European Food System Will Either Be Sustainable or Rapidly Become Unviable

The world of agriculture has to face up to a series of challenges, which are equally societal issues, around the question of food. These economic, strategic, environmental, climatic and human issues go beyond the scope of the farming profession. It is the role and responsibility of the political authorities to take a clear stand on the direction in which society wishes to develop the agriculture and food of the future, since these bodies have the power to legislate and to provide financial support or taxation to help achieve societal objectives.

Europe needs to build on its assets, which are the excellence of its agricultural, food and gastronomic products, as well as a tangible heritage (landscapes, soils and terroirs, rural architecture) and intangible heritage (farming know-how, agronomic science, innovations, flagship agri-food businesses, traditional products, chefs, etc.) of great value. While developing and preserving these precious assets, we need to drastically reduce the impact of agricultural production on the environment and, more particularly, on biodiversity, itself a source of life. By the same token, climate change requires agriculture and food systems to make a vital effort to adapt, so that it can continue to grow crops and produce food, as affirmed by the heads of state and government at the COP 28 [5].

Because climate change and biodiversity loss are the biggest threats to food security, the direction to be taken by agriculture is without doubt that of sustainability and resilience. Legislative and budgetary means should support farmers to move confidently in this direction. Laissez faire liberalism fosters doubt, uncertainty and market volatility. There has never been a "green revolution" in agriculture in this context. There is always government impetus and support when agriculture develops strongly, as history shows.

The best farming model with the capacity to simultaneously bring about economic performance, environmental protection and social equity is agro-ecology. As this term may designate a scientific discipline, a social movement or a set of agricultural techniques based on agronomic science, we need to define what we mean here by agro-ecology. We retain the definition given in a French agricultural law of 2014: "Agro-ecological production systems combine economic and social performance, in particular through a high level of social, environmental and health protection. These systems prioritise the autonomy of farms and the improvement of their competitiveness, by maintaining or increasing economic profitability, improving the added value of production and reducing the consumption of energy, water, fertilisers, plant protection products and veterinary medicines, in particular antibiotics. They are based on biological interactions and the use of ecosystem services, and the potential offered by natural resources, in particular water resources, biodiversity, photosynthesis, soil and air, while maintaining their capacity for renewal in terms of quality and quantity. They contribute to mitigating and adapting to the effects of climate change". The most successful agro-ecological systems are those developed by organic farming, permaculture, agroforestry, conservation agriculture, and regenerative farming, as well as several other environmental-friendly agricultural practices.

5. The EU Should Radically Turn Its CAP into a Transformative Policy Supporting Farm and Food Innovations

The CAP should evolve more rapidly in the direction of sustainability, which means rethinking the productivity model of development and opting for a radical shift to agro-ecology. Farmers would move beyond their role as primary producers of agricultural products and assume that of suppliers of public goods. This must be enshrined in European regulations at the highest level, as well as in a governance model that attributes to the regions the full capacity to finance and implement the optimal implementation of this agro-ecological policy.

Our central argument for turning the CAP into a transformative policy consists in radically reforming the way it delivers its support, together with other actions. As seen above, the 1st pillar eventually supports the continuation of unchanged agricultural practices, when we need investments to finance the agro-ecological transition. Rent-seeking mechanisms should be removed and the purpose of EU public aid to agriculture should be reconsidered, so that it encourages investment rather than supplementing farms' cash flow. The so-called 2nd pillar of the CAP, also known as the Rural Development branch of that policy, is precisely designed for financing physical and intellectual investment, for supporting environmental and climatic measures and helping the results of research and innovation to flourish among farms. Yet only a quarter of the CAP budget goes to the 2nd pillar. This should become more important, to the detriment of the first pillar, i.e., direct aid.

The new CAP would no longer count two pillars, associated with two different financial regimes (the EU budget pays in full the expenditure for the 1st pillar, and the 2nd pillar expenditure is co-financed between the European Commission and the Member States) but support one single budget regime, entirely co-financed, and targeting sustainability, climate adaptation, transformative innovation and agro-ecology. Hence, a sizeable source of funding will serve the transition. Furthermore, sharing the financial burden will increase the responsibility of all actors, while one single regime will mean simplification for national and regional administrations.

Other actions should envisage a debt relief programme for farmers, a special programme for favouring the entry of young people and women in the sector, a specific approach to agriculture in the Free Trade Agreements, a remuneration for carbon storage in soils, a ban on cages in animal rearing, a reconnection of animals and crops on the same farm, budgetary and fiscal support to organic food and farming, using genome-editing techniques with all due caution, a sufficient budget for research and innovation, openness to bottom-up initiatives, on-farm employment, free training and advice, and many more other actions. Each of these would merit further study and development.

At the other end of the food chain, consumer needs must be taken on board, not only in food pricing terms, but also in nutritional and health aspects. It is only by encompassing whole food systems that the transition to sustainability will succeed. The Farm to Fork strategy has opened up a promising avenue in this direction, which needs to be pursued with determination [6].

Funding: This research received no external funding.

Institutional Review Board Statement: Not applicable.

Informed Consent Statement: Not applicable.

Data Availability Statement: Data available from public domain resources.

Conflicts of Interest: The author declares no conflict of interest.

References

1. European Commission. *Communication from the Commission to the European Parliament, the Council, the European Economic and Social Committee and the Committee of the Regions—A Farm to Fork Strategy for a Fair, Healthy and Environmentally-Friendly Food System, COM(2020) 381 Final*; European Commission: Brussels, Belgium, 2020.
2. Eurostat. Agricultural Trade Statistics. 2023. Available online: https://ec.europa.eu/eurostat/web/products-eurostat-news/w/ddn-20230503-1 (accessed on 12 April 2024).
3. European Environmental Agency. *State Of Environment Report 2020—The European Environment, State and Outlook 2020, Knowledge for Transition to a Sustainable Europe*; European Environmental Agency: Copenhagen, Denmark, 2019.
4. European Commission. Direct Payments to Agricultural Producers, Graphs and Figures (Page 8). 2021. Available online: https://agriculture.ec.europa.eu/system/files/2023-03/direct-aid-report-2021_en.pdf (accessed on 12 April 2024).
5. COP28 Declaration. Available online: https://www.cop28.com/en/food-and-agriculture (accessed on 10 April 2024).
6. Bradley Harriet. Sustaining Tomorrow: The Imperative for the EU to Uphold its 2030 Goals for a Sustainable Food System. 2024. Available online: https://eu.boell.org/en/2024/03/20/eu-sustainable-food-system (accessed on 20 April 2024).

proceedings

MDPI

Proceeding Paper

Empowering Rural Communities on Rural Pact Implementation: A Human–Ecological Perspective on Social Innovation and Rural Young Entrepreneurship †

Maria João Parreira [1,*] and Iva Pires [1,2]

1 Research Group 4 Sustainability and Socio-Ecological Systems, Interdisciplinary Centre of Social Sciences (CICS.NOVA), 1099-085 Lisbon, Portugal
2 School of Social Sciences and Humanities (NOVA FCSH), 1069-061 Lisbon, Portugal; im.pires@fcsh.unl.pt
* Correspondence: mariaparreira@fcsh.unl.pt

† Presented at the XVIII International Seminars on Overarching Issues of the European Area, Porto, Portugal, 23–26 May 2024.

Abstract: This study aligns with the Rural Pact's Long-Term Vision for European Rural Areas by encouraging multi-level collaboration and stakeholder engagement to address rural needs. Using a Human Ecology perspective, it challenges rural stereotypes, promotes spatial justice and aims to reduce rural–urban disparities by fostering social innovation and youth entrepreneurship in rural areas. A thematic analysis of semi-structured interviews identified six key dimensions and their sub-dimensions within a transformative social–ecological model. The notable sub-dimensions include "Local Experimentation and Creativity" in Civil Society, "Higher Education Institutions and Lifelong Learning" in Knowledge and Skills, "Formal Institutional Monitoring at the National Level" in a Political and Institutional Context, and "Profiles—Young Age and Gender" in Entrepreneurship.

Keywords: human ecology; rural communities' empowerment; rural pact; rural young entrepreneur; social innovation; social–ecological systems; youth policies

1. Introduction

The European Union Rural Pact aims to amplify rural voices, elevate them on the political agenda, foster networking and mutual learning, and support voluntary commitments to advance the Long-Term Vision for European Rural Areas. This paper presents the preliminary results from ongoing doctoral research on Human Ecology, which frames itself as an "operational post-normal science" [1] (p. 13) to create a social–ecological model to evaluate and operationalize social innovation's role in boosting rural young entrepreneurship in disadvantaged areas. It emphasizes the new legal figure of the "Rural Young Entrepreneur" (Decree-Law n° 9/2019) and shares the vision of the recent OECD report "Assessing the Framework Conditions for Social Innovation in Rural Areas", which underscores the need for innovative approaches to address rural challenges [2]. Social innovation, a complex term with over 200 definitions, generally includes five elements: social needs, innovative solutions, implementation, improvements, and collaborations [3]. We use the European definition of social innovation employed by the Social Innovation in Marginalised Rural Areas (SIMRAs) project, which involves "reconfiguring social practices in response to societal challenges, enhancing societal well-being with civil society engagement" along with that of [4], which considers social and political capacities for citizen-led innovation. The key criteria for rural innovation investigated in this study include knowledge exchange, place-based policies, and stakeholder influence, addressing the question: how can the relevance of social innovation be assessed and applied to rural development, specifically to support young entrepreneurs in disadvantaged areas? The role of the "Rural Young Entrepreneur" in this context is further analyzed. Building on Nordberg's [5] perspective on social innovation and spatial justice, this study views the "capabilities of places" as assets, advocating for opportunities unique to each territory beyond equal resource distribution. Recognizing the diversity within rural areas, tailored interventions [6] and place-based strategies are essential for youth integration, economic diversity, and digital connectivity [7]. This paper presents the preliminary findings from stakeholder interviews conducted in person and online. A thematic analysis revealed six core dimensions (deductively coded) and sub-dimensions (abductively coded), forming part of a developing socio-ecological model that promotes a less anthropocentric approach to social innovation. The model, called RuSTIC—Rural Social–ecological Transformative Innovation Capabilities—draws inspiration from Huntjens' concept of Transformative Social–Ecological Innovation (TSEI) [8].

2. Materials and Methods

This study included 46 participants across four categories: young entrepreneurs, professors/researchers, NGOs and local associations, and government entities focused on social innovation and rural development. The participants were chosen through convenience sampling to capture a diversity of experiences across all of mainland Portugal including the NUTS 2 regions (pre-2024). The semi-structured interviews, held in person or via Zoom and lasting about 1.5 hour each, explored views on social innovation, rural young entrepreneurship, the challenges to rural development, resource diversity, and the role of Higher Education Institutions (HEIs) in promoting social innovation. The participants also discussed the Portuguese "Rural Young Entrepreneur" statute's potential to foster social innovation, territorial cohesion, and well-being in rural areas. The interviews were transcribed using TRINT software (free version), then were manually reviewed for accuracy and analyzed via MAXQDA through a thematic content analysis. So far, 24 interviews (5 young entrepreneurs, 7 professors/researchers, 7 NGOs, and associations, and 5 government entities) have been analyzed, resulting in approximately 700 pages of transcripts and around 4000 coded segments. The coding combined deductive elements from the European Innovation Scoreboard report "How to measure social innovation" [9] with abductive coding for the new sub-dimensions inspired by [10,11], yielding insights from both the established frameworks and empirical findings.

3. Results and Discussion

In Figure 1, we present the six main dimensions and their observed frequencies across all the coded segments. These dimensions encompass a total of 3904 coded segments, representing 99% of the relative frequencies from the 24 interviews coded so far. The remaining codes, outside of these main dimensions, such as "religion" and "critical mass", are less frequent. Although minimal in occurrence, these codes are acknowledged for their broader relevance to this study.

Among the six identified dimensions, Civil Society stands out, underscoring the importance of civil society participation as defined in the SIMRA project. In each of the four most prominent dimensions, key sub-dimensions emerged including the following: "Local Experimentation and Creativity" for Civil Society; "Higher Education Institutions and Lifelong Learning" for Knowledge and Skills; "Formal Institutional Monitoring at National Level" for a Political and Institutional Context; and "Young Age and Gender Profiles" for Entrepreneurship. The preliminary results emphasize the essential role of civil society in "Local Experimentation and Creativity", which is supported by higher education, institutional monitoring, and diverse youth profiles in entrepreneurship. The findings highlight the need for tailored support for young entrepreneurs, collaborative approaches, and fewer bureaucratic barriers. The key factors for promoting social innovation include public policies, infrastructure, political support, sustainable agriculture, and adaptable funding. This includes integrating the knowledge, skills, and technologies, especially digital tools, that can attract young people to rural areas and enable remote work. Education and training, as well as partnerships among universities, businesses, and social organizations, are crucial for practical solutions and improving rural communities' quality of life. Trust between civil society, policymakers, higher education institutions, and support agencies is vital for a sustainable transformation. Social innovation must also address the underlying power dynamics to prevent local elite capture [12].

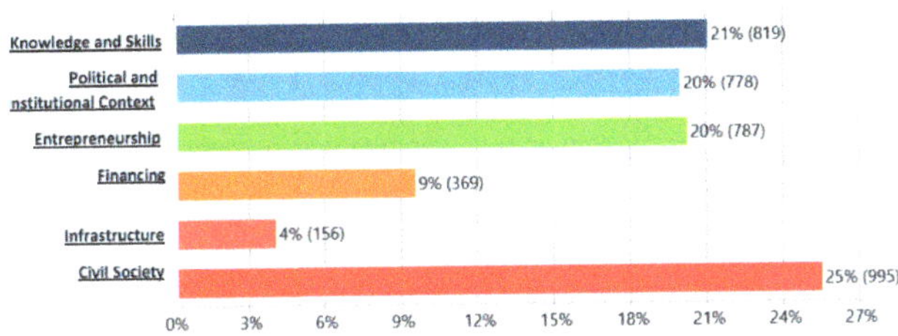

Figure 1. Elaboration performed in MAXQDA 24 Release 24.3.0 for this study: Six Main Dimensions and Their Frequencies in Coded Segments.

4. Main Considerations and Future Actions

A collaborative network of critical academics, "organic" intellectuals, reflective activists, and progressive policymakers working closely with local communities is essential to address the rural challenges in Europe effectively, beyond single theories or temporary ideas [13]. Addressing rural issues requires integrating urban and rural development, utilizing regional strengths, and empowering local actors, as exclusively rural strategies are insufficient in a post-urban context [14]. Furthermore, local rural sustainability initiatives should prioritize social equity over mere economic growth and establish inclusive, locally relevant goals [15]. This research aims to inform youth policies, keeping rural areas central in the political arena and empowering young rural entrepreneurs to lead sustainable, innovative community development aligned with the European Union Rural Pact.

Author Contributions: M.J.P.—conceptualization, methodology, software use, and writing—original draft preparation, review, and editing; I.P.—review, editing, and supervision. All authors have read and agreed to the published version of the manuscript.

Funding: This research is being funded by National Funds through FCT, the Portuguese Foundation of Science and Technology, under the Ph.D. fellowship 2021.06974.BD: https://doi.org/10.54499/2021.06974.BD.

Institutional Review Board Statement: The study is being conducted following the Declaration of Helsinki. Although no sensitive data is being processed, approval has been requested from the FCSH-NOVA Ethics Committee, and the respective approval code is pending.

Informed Consent Statement: Informed consent was obtained from all the interviewees involved in this study.

Data Availability Statement: Not applicable at this phase. When all the data has been processed it will be shared including the links to the open-access archived datasets generated during this study/project.

Conflicts of Interest: The authors declare no conflicts of interest. The funders had no role in the design of this study; in the collection, analyses, or interpretation of data; in the writing of the manuscript; or in the decision to publish the results.

References

1. Dyball, R. A brief history of human ecology within the Ecological Society of America and speculation on future direction. *Hum. Ecol. Rev.* **2017**, *23*, 7–16. [CrossRef]
2. Organization for Economic Co-Operation and Development (OECD). Assessing the Framework Conditions for Social Innovation in Rural Areas. OECD Local Econ. Employment Dev. (LEED) Pap. OECD Publishing. 2024. Available online: https://www.oecd.org/en/publications/assessing-the-framework-conditions-for-social-innovation-in-rural-areas_74367d76-en.html (accessed on 20 May 2024).
3. Eichler, G.M.; Schwarz, E.J. What sustainable development goals do social innovations address? A systematic review and content analysis of social innovation literature. *Sustainability* **2019**, *11*, 522. [CrossRef]
4. Moulaert, F. Social Innovation: Institutionally Embedded, Territorially (Re)Produced. In *Social Innovation and Territorial Development*; MacCallum, D., Moulaert, F., Hillier, J., Vicari Haddock, S., Eds.; Routledge: New York, NY, USA, 2016; eBook; pp. 27–40. ISBN 9781315609478.
5. Nordberg, K. Spatial justice and local capability in rural areas. *J. Rural Stud.* **2020**, *78*, 47–58. [CrossRef]
6. Ferrão, J. Despovoamento em áreas rurais: Entre a inevitabilidade e a capacidade de transformação. *Cultivar. Cad. Anál. Prosp.* **2018**, *11*, 13–19.
7. Proietti, P.; Sulis, P.; Perpiña Castillo, C.; Lavalle, C.; Aurambout, J.P.; Batista E Silva, F.; Bosco, C.; Fioretti, C.; Guzzo, F.; Jacobs, C.; et al. *New Perspectives on Territorial Disparities*; EUR 31025 EN; Publications Office of the European Union: Luxembourg, 2022; ISBN 978-92-76-49485-0. JRC126033. [CrossRef]
8. Huntjens, P. *Towards a Natural Social Contract: Transformative Social-Ecological Innovation for a Sustainable, Healthy and Just Society*; Springer Nature: Berlin/Heidelberg, Germany, 2021; p. 205. [CrossRef]
9. European Innovation Scoreboards (EIS) Project. How to Measure Social Innovation. Exploratory Report. European Commission. 2021. Available online: https://ec.europa.eu/docsroom/documents/45665/attachments/1/translations/en/renditions/native (accessed on 13 April 2021).
10. Thompson, J. A Guide to Abductive Thematic Analysis. *Qual. Rep.* **2022**, *27*, 1410–1421. [CrossRef]
11. Vila-Henninger, L.; Dupuy, C.; Van Ingelgom, V.; Caprioli, M.; Teuber, F.; Pennetreau, D.; Bussi, M.; Le Gall, C. Abductive coding: Theory building and qualitative (re) analysis. *Sociol. Methods Res.* **2024**, *53*, 968–1001. [CrossRef]
12. Slee, B.; Lukesch, R.; Ravazzoli, E. Social innovation: The promise and the reality in marginalised rural areas in Europe. *World* **2022**, *3*, 237–259. [CrossRef]
13. Ferrão, J. Animar localmente prefigurações de futuros desejados. Da inovação social à transformação societal. In *Entre Urgências e Utopia: Múltiplas Escalas da ação: Que Mundo Estamos a Construir?* Consequência Editora: Rio de Janeiro, Brasil, 2023; pp. 37–62.
14. Westlund, H.; Borsekova, K. Rural problems, policies and possibilities in a post-urban world. *Reg. Sci. Policy Pract.* **2023**, *15*, 717–728. [CrossRef]
15. Mahon, M.; Woods, M.; Farrell, M.; Jones, R.; Goodwin-Hawkins, B. A spatial justice perspective on EU rural sustainability as territorial cohesion. *Sociol. Ruralis* **2023**, *63*, 683–702. [CrossRef]

proceedings

Proceeding Paper

Challenges for Territorial Sustainability After Forest Fires: The Case Study of the Guarda District †

Elisabete Soares [1,2,*], Fátima David [1,3] and Pedro M. S. M. Rodrigues [1,4]

1 Polytechnic of Guarda, 6300-559 Guarda, Portugal; sdavid@ipg.pt (F.D.); prodrigues@ipg.pt (P.M.S.M.R.)
2 Technology, Restoration and Arts Enhancement Center, Polytechnic of Tomar, 2300-313 Tomar, Portugal
3 Research Centre on Accounting and Taxation, School of Management, Polytechnic of Cávado and Ave, 4750-810 Barcelos, Portugal
4 Chemistry Research Unit (CIQUP), IMS–DGAOT, Faculty of Sciences of University of Porto, 4169-007 Porto, Portugal
* Correspondence: esoares@ipg.pt
† Presented at the XVIII International Seminars on Overarching Issues of the European Area, Porto, Portugal, 23–26 May 2024.

Abstract: Forest fires cause significant harm to ecological communities and society by altering land cover changes, imposing economic burdens, and exacerbating global warming and climate change. Based on burned areas and their land cover, this study analyzes the burned areas in Portugal's Guarda District. It examines the environmental and economic impacts of the forest fires occurring in August 2022. The results highlight that 9.42% of the Pinus Sylvestris, a native species of the Serra da Estrela Natural Park, was lost, alongside other ecological damage, including a decrease in greenhouse gas (GHG) sequestration capacity and soil degradation, which increases the risk of erosion and reduces the fertility of the affected areas. The economic impact was also substantial, with significant costs related to firefighting efforts, loss of timber resources, and adverse effects on tourism and local livelihoods.

Keywords: forest fires; greenhouse gas emissions; territorial sustainability; land cover change

1. Introduction

Forests and soil are resilient and crucial carbon reservoirs. They absorb significant carbon dioxide (CO_2) emissions, contributing to the global carbon cycle. Forests, with their microclimate regulation, influence temperature and humidity, and are essential for mitigating climate change's harmful effects [1]. In the summer, temperature differences can increase by up to 10 °C between forests and adjacent urban areas [1,2]. The impact of such forest fires is not limited to the environment, but also extends to substantial economic and social repercussions. Despite the challenges, forests continue to thrive, occupying 43.5% of Europe's land area, increasing by almost 10% between 1990 and 2020 [3]. In Portugal, forests are the primary land use, covering 36% of the territory [4]. Forests, having the ability to act as carbon sources or carbon sinks, are a testament to their resilience. On the one hand, forest carbon is released when trees burn or decay after dying, but on the other hand, forest is also a carbon sink if it absorbs more carbon from the atmosphere through photosynthesis than it releases [5]. In the world's forests, over a trillion tons of CO_2 is stored in biomass [6], and EU forests absorb approximately 10% of the EU's annual greenhouse gas (GHG) emissions [1,3], highlighting their role in mitigating climate change [7]. The amount of additional carbon that can be stored in forests depends on (1) how much society is willing to spend to store more carbon and (2) how quickly the carbon must be stored,

https://doi.org/10.3390/proceedings2025113003

i.e., the faster the carbon is stored, the more expensive it will be [6]. Carbon sinks play an essential role in the transition to climate neutrality in the EU [8], and, in particular, the agriculture, forestry, and land use sectors can make an essential contribution in that context. The Land Use, Land Use Change, and Forestry (LULUCF) sector, acting as a carbon sink, is a central instrument to the EU's overarching objective to reduce GHG emissions by 55% to 57% compared to 1990 levels and achieve carbon neutrality by 2050 [8]. According to Regulation (EU) 2023/839 [8], each Member State in the LULUCF sector must set targets to increase GHG removals during the period 2026–2030, resulting in a target of 310 million of CO_2 equivalent of net removals for the EU as a whole in 2030, which would further contribute to achieving the Union's climate objectives. Forest fires also contribute to climate change, releasing substantial amounts of CO_2 and other GHG, intensifying temperature rises and humidity decreases, and leading to more frequent and intense fires, compromising the carbon neutrality targets for 2050 [8]. The total burned area in Continental Portugal between 1st January and 31st August 2022, was 106,639 ha, 36% more than the annual average for 2012–2021 [9]. According to [9], 2022, up to August 31st, was the year with the sixth-highest number of fires and the fourth-highest value in burned area since 2012. The resultant decline in Portugal's forest area from fires could jeopardize the EU's 2030 goal to increase carbon sinks by 15%. Moreover, forest fires have far-reaching implications for territorial sustainability, causing significant changes in land cover and soil structure, water quality, geological hazards such as debris flows, landslides, and rockfalls, and post-fire hydrological events, among others [10], profoundly affecting ecological, economic, and social systems [11–13]. Supported by a literature review and using data on burned areas and their land cover, this study seeks to provide insights into the consequences of forest fires in the region. Thus, it presents a brief analysis of economic and environmental impacts resulting from forest fires in the Guarda District in August 2022.

2. Materials and Methods

2.1. Study Area

The emphasis of the analysis is placed on the Guarda District (Figure 1), whose territory suffered serious damage from the fires that occurred in August 2022. Guarda District is a territory with 5535.31 km^2, with mountainous relief, including part of Serra da Estrela Natural Park, and has a population density of 25.83 inhabitants per km^2.

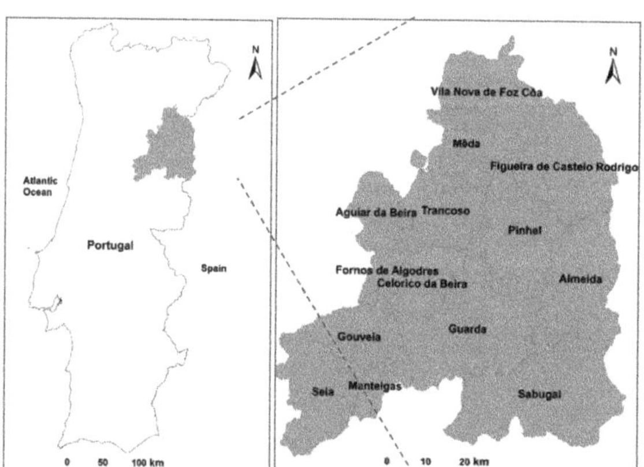

(a) (b)

Figure 1. Study area. Geographical framework of the Guarda District (highlighted in gray) in Portugal (**a**); municipalities of Guarda District (**b**).

2.2. Datasets

2.2.1. Land Cover

Land cover classes were analyzed, based on the Map of Land Use and Land Cover 2018 [14], to understand the changes in land cover and the impacts of forest fires in 2022. Accordingly, in 2018, the classes of wilds and forests prevailed with 334,192.13 ha, corresponding to 60.4% of the territory—176,654.330 ha (31.9%) by the forest class and 157,537.83 ha (28.5%) by the wild class. The resinous forest class occupied 15.1% (83,729.76 ha), distributed by 14% (77,362.66 ha) *Pinus pinaster*, 0.1% (476.49 ha) *Pinus pinea*, and 1% (5890.61 ha) other resinous (Pinus Sylvestris) classes. The hardwood forest (oaks, chestnut trees, holm oaks, cork oaks, invasive) class covered 92,924.54 ha (16.8%), and Eucalyptus only 0.6% (3383.41 ha) of the district area.

2.2.2. Burned Area/Land Cover

In August 2022, a large fire engulfed approximately 28,000 ha of the Serra da Estrela. Between 2nd and 22nd August 2022, 20,618 ha of the Guarda District area was scorched [12], causing significant changes in land cover. This area is a habitat for several protected autochthonous species, whose environment, once damaged by fires, often faces irreparable harm. Hence, the period from 2nd to 22nd August was selected for our analysis. Table 1 details the burned area by land cover type, focusing on forests and wilds classes.

Table 1. Burned area in Guarda District by forest class and wild cover from 2nd to 22nd August 2022.

Land Cover Classes	Burned Area (ha)	Percentage (%)
Pinus pinaster	2970.42	14.41
Pinus pinea	1.33	0.006
Other resinous	554.83	2.69
Hardwood forest	3213.62	15.59
Wilds	8741.95	42.40
Eucalyptus	662.19	3.21
Total	16,144.34	78.31

Source: [15].

3. Results and Discussion

Considering Table 1, 78.31% of the burned area in August 2022 was forests and wilds: 7402.39 ha (35.90%) of forest and 8741.95 ha (42.40%) of wild. The hardwood forest, with 3213.62 ha, corresponds to 15.59% of the area affected by wildfires. Also, the *Pinus pinaster* had a significant burned area (2970.42 ha, i.e., 14.41%). The other resinous and eucalyptus classes burned 2.69% and 3.21%, respectively, and *Pinus pinea* was the class with the smallest burned area, 1.33 ha (0.006%). Considering the land cover area of each class in 2018, the following were burned: 3.84% of *Pinus pinaster*, 0.28% of *Pinus pinea*, 9.42% of other resinous, 3.46% of hardwood forest, 5.55% of wilds, and 19.57% of eucalyptus. The Pinus Sylvestris, classified as other resinous, is an autochthonous species from Serra da Estrela Natural Park, which is threatened and at risk of extinction, highlighting its small area (1% in 2018) of occupation in the district. The *Pinus pinea*, grown for its seed (pine nuts), also represents an environmental and economic loss, even the tiny burned area. Wood is used for several requests, with 3.84% of its burned area having negative economic impacts. Likewise, the 19.57% eucalyptus burned area presents economic losses. In addition, forest fires have a significant environmental impact, not only for releasing amounts of GHG but also due to the nature that dies and the loss of forests that no longer contribute to carbon sequestration and sinking. It is estimated that a hectare of trees can contain around 50 tons of carbon, approximately 180 tons of CO_2 in the atmosphere [16], and sequester 1 to 2 tons of CO_2

annually [17]. Some forests can store up to 10 tons of carbon per hectare, while others store much more, but according to these authors [16], the estimate of 50 tons per hectare would be typical for a young forest. Thus, it is legitimately estimated that 7402.39 ha of burned forest stopped storing 370,119.50 tons of carbon and sequestering 7402.39 to 14,804.78 tons of CO_2 from the atmosphere per year. The burned wild area also stopped contributing to carbon absorption.

4. Conclusions

The 16,144.34 ha forest and wild areas burned in the Guarda District in August 2022 caused 9.42% of the Pinus Sylvestris, an autochthonous species from Serra da Estrela Natural Park, to disappear, beyond other ecological damage, such as the contribution of forest species to the reduction in GHG emissions. It was estimated that the 7402.39 ha of burned forest could hurt the sequestering CO_2 of around 10,000 tons. Furthermore, other impacts occur at an economic and social level. Forest regeneration takes time, and full recovery is difficult, especially for native species. The decline in the population of plant species, particularly autochthonous species, is worrying, damaging the balance of habitats and ecosystems and the environment in general. Thus, this study provides some warnings about the consequences of forest fires that can present challenges to the sustainability of the territory, aspiring to facilitate a balanced coexistence between society and nature in this ecosystem.

Author Contributions: All authors (E.S., F.D. and P.M.S.M.R.) contributed equally to the paper. All authors have read and agreed to the published version of the manuscript.

Funding: This research received no external funding.

Institutional Review Board Statement: Not applicable.

Informed Consent Statement: Not applicable.

Data Availability Statement: All the data will be made available on request to the authors.

Conflicts of Interest: The authors declare no conflicts of interest.

References

1. EEA. European Forest Ecosystems: Key Allies in Sustainable Development. 2023. Available online: https://www.eea.europa.eu/publications/european-forest-ecosystems-key-allies (accessed on 23 March 2024).
2. Haesen, S.; Lembrechts, J.J.; Frenne, P.; Lenoir, J.; Aalto, J.; Ashcroft, M.B.; Kopecký, M.; Luoto, M.; Maclean, I.; Nijs, I.; et al. ForestTemp-Sub-canopy microclimate temperatures of European forests. *Glob. Chang. Biol.* **2021**, *28*, 7157–7158. [CrossRef] [PubMed]
3. European Parliament. Climate Change: Better Using EU Forests as Carbon Sinks. 2023. Available online: https://www.europarl.europa.eu/topics/en/article/20170711STO79506/climate-change-better-using-eu-forests-as-carbon-sinks (accessed on 11 June 2024).
4. ICNF. Portugal Perfil Florestal. 2021. Available online: https://www.icnf.pt/api/file/doc/1f924a3c0e4f7372 (accessed on 7 June 2024).
5. Natural Resources Canada. Available online: https://natural-resources.canada.ca/climate-change/climate-change-impacts-forests/forest-carbon/13085 (accessed on 12 June 2024).
6. Mendelsohn, R.; Sedjo, R.; Sohngen, B. Forest Carbon Sequestration. In *Fiscal Policy to Mitigate Climate Change: A Guide for Policymakers*; Parry, I.W.H., Mooij, R., Keen, M., Eds.; International Monetary Fund: Washington, DC, USA, 2012; Chapter 5, pp. 89–102.
7. Blumröder, J.; May, F.; Härdtle, W.; Ibisch, P.L. Forestry contributed to warming of forest ecosystems in northern Germany during the extreme summers of 2018 and 2019. *Ecol. Solut. Evid.* **2021**, *2*, 1–14. [CrossRef]
8. EU. Regulation (EU) 2023/839 of the European Parliament and of the Council of 19 April 2023. Available online: https://eur-lex.europa.eu/eli/reg/2023/839/oj (accessed on 12 June 2024).

9. DGPFR. *5.º Relatório Provisório de Incêndios Rurais de 2022*; 5.º RPIR/DGPFR/2022; Instituto da Conservação da Natureza e Florestas—Divisão de Gestão do Programa de Fogos Rurais: Lisboa, Portugal, 2022; p. 14.
10. USAID/USFS. Portugal Wildfires. Burned Area Emergency Response (BAER) Review. Report USAID/USFS 2022. Available online: https://www.icnf.pt/florestas/gfr/gfrgestaoinformacao/grfrelatorios/recuperacaodeareasardidas (accessed on 11 June 2024).
11. Lourenço, L. Forest fires in continental Portugal Result of profound alterations in society and territorial consequences. *Méditerranée* **2018**, *130*, 1–31. [CrossRef]
12. Castro, A.C.M.; Nunes, A.; Sousa, A.; Lourenço, L. Mapping the Causes of Forest Fires in Portugal by Clustering Analysis. *Geosciences* **2020**, *10*, 53. [CrossRef]
13. Ferreira, C.; Pinto, L.C.; Valente, M. Forest fire causes and prevention strategies in Portugal: Insights from stakeholder focus groups. *For. Policy Econ.* **2024**, *169*, 103330. [CrossRef]
14. DGT. *Especificações Técnicas da Carta de Uso e Ocupação do Solo (COS) de Portugal Continental para 2018*; Relatório Técnico; Direção-Geral do Território: Lisboa, Portugal, 2019.
15. Branco, L. Mapeamento de Incêndios e Caracterização Geral do Território a Partir de Deteção Remota. Bachelor's Thesis, Polytechnic of Guarda, Guarda, Portugal, December 2022. Available online: https://bdigital.ipg.pt/dspace/handle/10314/9521 (accessed on 23 March 2024).
16. Climate Portal. *How Many New Trees Would We Need to Offset Our Carbon Emissions?* MIT Climate Portal. Available online: https://climate.mit.edu/ask-mit/how-many-new-trees-would-we-need-offset-our-carbon-emissions (accessed on 23 March 2024).
17. Klaus, J. Forests Change the Climate. *Max Planck Res.* **2020**, *4*, 78–85.

proceedings

MDPI

Proceeding Paper

The Vulnerability of Tourism to Climate Change in Portuguese and Brazilian Cities—A Review †

Hélder Silva Lopes [1,2,3,*] and Diego Tarley Ferreira Nascimento [4]

1 Lab2PT—Laboratory of Landscapes, Heritage and Territory, Department of Geography, University of Minho, Campus de Azurém, Av. da Universidade, 4800-058 Guimarães, Portugal
2 IdRA—Water Research Institute, Group of Climatology, Department of Geography, University of Barcelona, Carrer de Montalegre, 6, Ciutat Vella, 08001 Barcelona, Spain
3 Composing Worlds Network, Fernando Pessoa Foundation, Praça 9 de Abril, 349, 4249-004 Porto, Portugal
4 Instituto de Estudos Socioambientais, Federal University of Goiás, Avenida Esperança s/n, Câmpus Samambaia—Prédio da Reitoria, Goiânia 74690-900, Goiás, Brazil; diego_nascimento@ufg.br
* Correspondence: helderlopes@geografia.uminho.pt; Tel.: +351-932-669-077
† Presented at the XVIII International Seminars on Overarching Issues of the European Area, Porto, Portugal, 23–26 May 2024.

Abstract: Tourism and climate change exhibit a binomial relationship, in which the former contributes to and is affected by the latter. In this context, the vulnerabilities of the tourism sector to climate change are assessed, focusing on cities in Portugal and Brazil, based on a literature review. A literature review was conducted using the keywords "tourism", "climate change", "Brazil", "Portugal", and their derivatives in Portuguese on Web of Science, Scopus, and Google Scholar databases. In both countries, extreme rainfall events cause delays or interrupt travel, and affect infrastructure, equipment, and services for tourists. In addition, heat waves and droughts have caused thermal discomfort and water shortages for residents and tourists. Adaptation and mitigation strategies are therefore urgently needed in both territories.

Keywords: tourism; global change; climate resilience; review

1. Introduction

Tourism and climate change have a binomial relationship, with tourism being both sensitive to and responsible for climate change [1,2]. Rising temperatures, rising sea levels, and the occurrence of extreme weather events have a negative impact on the tourism sector, reducing visitor numbers, altering visitor flows, fragmenting the length of stays, and increasing operating costs, especially for air conditioning in accommodation [2,3]. Nevertheless, tourism is responsible for about 5% of global greenhouse gas (GHG) emissions, with air transport being one of the main sources of emissions [4,5]. In this context, climate change is the greatest challenge for sustainable tourism in the 21st century [6].

Despite growing recognition of the need for sustainable tourism practices, significant vulnerabilities and knowledge gaps remain [7,8].

Therefore, this literature review assesses the vulnerability of the tourism sector to climate change, focusing on cities in Portugal and Brazil. The study is derived from an ongoing research project that aims to contribute to climate change resilience by proposing a local governance agenda for several cities in Brazil, Mozambique, Portugal, and Spain, funded by the Brazilian National Council for Scientific and Technological Development (CNPq).

Research on the impacts of climate change on tourism, particularly in urban areas, still has significant gaps that hinder effective adaptation and mitigation strategies. These gaps include a lack of comprehensive analyses across different regions, a limited understanding of the impacts of extreme weather events, and a lack of integrated assessments at the destination level [9]. In addition, there is an urgent need for decision-relevant information and tailored climate services [10].

2. Materials and Methods

To carry out the study, papers and other scientific works were consulted on the Web of Science, Scopus, and Google Scholar portals, using the keywords "tourism", "climate change", "Brazil", "Portugal", and their derivations in Portuguese, with the Boolean operator AND, and without restriction on the type of work or period of publication.

3. Results and Discussion

The Intergovernmental Panel on Climate Change's (IPCC) Sixth Assessment Report (AR6) paints a clear and alarming picture, at the 99% confidence level, that human activities have undeniably caused the planet's temperature to rise, leading to rapid and widespread changes in the atmosphere, oceans, ice sheets, and living systems [11].

Portugal is expected to experience a decline in tourism due to rising temperatures, making it one of the European countries with the greatest decline in tourism over the coming decades (third place), with a 5% decline in coastal regions [12].

The implementation of structural measures and effective collaborative planning for urban tourism adaptation to climate change by 2050 is essential, with the need for government and community guidance in defining adaptation proposals, especially in key areas for sustainable practices: energy-focused hotel solutions, green infrastructure development, and stakeholder network participation [13]. In addition, another important action is the ability to map risk areas, with an example of a methodology capable of assessing outdoor thermal comfort in the tourist city of Porto—Portugal [14]. Other major cities in Portugal are suffering from the worsening of extreme phenomena, such as heatwaves, which are becoming more frequent and intense (Figure 1A), or with flooding and waterlogging, affecting the city of Lisbon (Figure 1B).

Despite the studies presented for urban areas with higher temperatures exacerbated by Urban Heat Island (UHI), the literature reviewed indicates that sun and beach tourism in Portugal is particularly vulnerable to climate change [2,6–16], and this is also a major concern in the case of Brazil, where 69% of national and international tourist flows are directed to coastal regions [17–19].

The Brazilian cities with the highest tourist demand (e.g., Rio de Janeiro, Salvador, Recife, Natal, Florianópolis, and Balneário Camboriú) face aggravated challenges due to the greater frequency and intensity of extreme precipitation events that cause sudden sea inundations, floods, and landslides [17–20], as illustrated in Figure 1C.

(A)

(B)

Figure 1. *Cont.*

(C) (D)

Figure 1. Some evidence of climate change in tourism sector in Portugal and Brazil. (**A**) Heat waves during the summer in Lisbon—Tower of Belém. Source: CNN Portugal (2024) [21]. (**B**) Flooding in a central area illustrates the risks of climate change. Source: A Mensagem (2022) [22]. (**C**) Waves invade the shore of Rio de Janeiro (RJ/Brazil) during a sea rebound on 6 November 2023. Source: O Globo (2023) [23]. (**D**) Porto Alegre Airport (RS/Brazil) closed after the flooding of Lake Guaíba in May 2024. Source: CNN [24].

In general, in both the Portuguese and Brazilian cases, adverse climatic events can cause disruptions and closures of roads and airports, delaying or interrupting travel [2,3] and affecting infrastructure, facilities, destinations, and services for tourists [5,20,25–29]. In addition, heat waves and droughts are becoming more frequent and intense, causing thermal discomfort and water shortages for residents and tourists [3,5,18,28].

4. Conclusions

Since climate change is no longer a future scenario, but a daily reality on a global scale, adaptation and mitigation strategies are urgently needed to protect local populations and the tourism sector. Some measures have been implemented in European cities, such as the strategy of adaptive architecture and climate refuges for tourists, already implemented in Porto [21], and with the prospect of being adopted in Lisbon. However, mitigation and adaptation actions and measures are still in the early stages of development in Brazil, especially those that consider the tourism sector, activities, and destinations.

Author Contributions: Conceptualization, H.S.L. and D.T.F.N.; methodology, D.T.F.N.; investigation, H.S.L. and D.T.F.N.; resources, H.S.L. and D.T.F.N.; writing—original draft preparation, H.S.L. and D.T.F.N.; writing—review and editing, H.S.L.; project administration, D.T.F.N.; funding acquisition, D.T.F.N. All authors have read and agreed to the published version of the manuscript.

Funding: This research was funded by Brazil's National Council for Scientific and Technological Development, grant number 443297/2023-1 and 305318/2022-5.

Institutional Review Board Statement: Not applicable.

Informed Consent Statement: Not applicable.

Data Availability Statement: Data sharing is not applicable to this article.

Conflicts of Interest: The authors declare no conflicts of interest and the funder had no role in the design of the study; in the collection, analyses, or interpretation of data; in the writing of the manuscript; or in the decision to publish the results.

References

1. Gössling, S.; Peeters, P.; Ceron, J.P.; Dubois, G.; Patterson, T.; Richardson, R.B. The eco-efficiency of tourism. *Ecol. Econ.* **2005**, *54*, 417–434. [CrossRef]
2. Manchete, R. Clima e turismo num contexto de mudanças climáticas. *Finisterra* **2011**, *91*, 139–154.

3. Grimm, I.J.; Prado, L.; Giacomitti, R.B.; Mendonça, F. Mudanças climáticas e o turismo: Desafios e possibilidades. *Rev. Bras. Climatol.* **2012**, *11*, 55–78. [CrossRef]
4. Simpson, M.C.; Gössling, S.; Scoot, D.; Hall, C.; Gladin, E. *Climate Change Adaptation and Mitigation in the Tourism Sector: Frameworks, Tools and Practices*; UNEP: Nairobi, Namibia; University of Oxford: Oxford, UK; UNWTO: Madrid, Spain; WMO: Paris, France, 2008.
5. Gil, J.; Marques, N.R.; Andrade, G.N. Agenda climática e o turismo no Brasil: Contribuições para políticas públicas de adaptação frente às mudanças climáticas. *Rev. Bras. Pesqui. Tur.* **2023**, *17*, 2759. [CrossRef]
6. Scott, D. Why sustainable tourism must address climate change. *J. Sustain. Tour.* **2011**, *19*, 17–34. [CrossRef]
7. Scott, D.; Hall, C.M.; Gössling, S. A review of the IPCC Fifth Assessment and implications for tourism sector climate resilience and decarbonization. *J. Sustain. Tour.* **2019**, *24*, 8–30. [CrossRef]
8. Gössling, S.; Higham, J. The Low-Carbon Imperative: Destination Management under Urgent Climate Change. *J. Travel Res.* **2021**, *60*, 1167–1179. [CrossRef]
9. Lopes, H.S.; Remoaldo, P.; Silva, M.; Ribeiro, V.; Vide, J.M. Climate in tourism's research agenda: Future directions based on literature review. *Boletín Asoc. Geógrafos Españoles* **2021**, *90*, 1–58. [CrossRef]
10. Remoaldo, P.; Lopes, H.S.; Ribeiro, V.; Alves, J.A. *Tourism and Climate Change in the 21st Century: Challenges and Solutions*; Advances in Spatial Science; Springer: Cham, Switzerland, 2024.
11. IPCC. *IPCC Sixth Assessment Report (AR6): Climate Change 2022—Impacts, Adaptation and Vunerability*; International Panel for Climate Change: Geneve, Switzerland, 2023.
12. European Comission. *Regional Impact of Climate Change on European Tourism Demand*; Publications Office of the European Union: Luxemburg, 2023. [CrossRef]
13. Lopes, H.S.; Remoaldo, P.C.; Ribeiro, V.; Martin-Vide, J. Pathways for adapting tourism to climate change in an urban destination—Evidences based on thermal conditions for the Porto Metropolitan Area (Portugal). *J. Environ. Manag.* **2022**, *315*, 115161. [CrossRef] [PubMed]
14. Lopes, H.S.; Remoaldo, P.C.; Ribeiro, V.; Martín-Vide, J. A comprehensive methodology for assessing outdoor thermal comfort in touristic city of Porto (Portugal). *Urban Clim.* **2022**, *45*, 101264. [CrossRef]
15. Perch-Nielsen, S.L. The vulnerability of beach tourism to climate change: An index approach. *Clim. Chang.* **2010**, *100*, 579–606. [CrossRef]
16. Scott, D.; Verkoeyen, S. Assessing the climate change risk of a coastal-island destination. In *Global Climate Change and Coastal Tourism: Recognizing Problems, Managing Solutions and Future Expectations*; Jones, A., Phillips, M., Eds.; Cabi Intenational: Oxford, UK, 2017; pp. 62–73.
17. Brandão, C.N.; Barbieri, J.C.; Araújo, M.F. Tourism and climate: A substantive theory on adaptation strategies in the accommodation sector in "águas paulistas" circuit, São Paulo, Brazil. *WIT Trans. Ecol. Environ.* **2018**, *227*, 163–174.
18. Grimm, I.J. Impactos das mudanças climáticas no sistema turístico: O caso brasileiro. *Cad. Virtual Tur.* **2019**, *19*, 1–14. [CrossRef]
19. Brazil. *Anuário Estatístico do Turismo 2018*; Ministério do Turismo: Brasília, Brazil, 2019.
20. Moreno, A.; Becken, S. A climate change vulnerability assessment methodology for coastal tourism. *J. Sustain. Tour.* **2009**, *17*, 473–488. [CrossRef]
21. CNN Portugal. Alterações Climáticas Podem Reduzir em Seis Meses Esperança Média de Vida Humana. Available online: https://cnnportugal.iol.pt/alteracoes-climaticas/esperanca-media-de-vida/alteracoes-climaticas-podem-reduzir-em-seis-meses-esperanca-media-de-vida-humana/20240118/65a976a8d34e65afa2f9cf71 (accessed on 5 June 2024).
22. A Mensagem. Inundações. "Área de Lisboa dá Péssimo Exemplo a Evitar Riscos das Alterações Climáticas—Sobretudo na Zona Ribeirinha". Available online: https://amensagem.pt/2022/12/13/inundacoes-chuva-lisboa-area-metropolitana-exemplo-sergio-barroso-cheias-riscos-alteracoes-climaticas/ (accessed on 5 June 2024).
23. O Globo. Ressaca faz Mar Invadir Calçadão na Zona Sul do Rio. 2023. Available online: https://oglobo.globo.com/rio/noticia/2023/11/05/ressaca-faz-mar-invadir-calcadao-na-zona-sul-do-rio-veja-video.ghtml (accessed on 10 June 2024).
24. CNN Brasil. Enchente no RS: Novas Imagens Aéreas Mostram Como Está a Cidade de Porto Alegre. 2024. Available online: https://www.cnnbrasil.com.br/nacional/enchente-no-rs-novas-imagens-aereas-mostram-como-esta-a-cidade-de-porto-alegre-veja/ (accessed on 10 June 2024).
25. Fang, Y.; Yin, J.; Wu, B. Climate change and tourism: A scientometric analysis using CiteSpace. *J. Sustain. Tour.* **2017**, *26*, 108–126. [CrossRef]
26. Marengo, J.A.; Muller-Karger, F.; Pelling, M.; Reynolds, C.; Merrill, S.B.; Nunes, L.H.; Paterson, S.; Gray, A.J.; Lockman, J.; Kartez, J.; et al. A globally deployable strategy for co-development of adaptation preferences to sea-level rise: The public participation case of Santos, Brazil. *Nat. Hazards* **2017**, *88*, 39–53. [CrossRef]
27. Martellotta, A.; Lobo, H. As Mudanças Climáticas Globais deveriam incomodar os Resorts Litorâneos do Nordeste Brasileiro? *Rev. Tur. Estud. Práticas* **2023**, *12*, 1–18.

28. Santos, E. da S.; Marengo, J.A. Desafío e impacto del cambio climático em el turismo: El escenario brasileño. *Estud. Perspect. Tur.* **2020**, *29*, 864–885.
29. Martín, M.B.G. Weather, climate and tourism a geographical perspective. *Ann. Tour. Res.* **2005**, *32*, 571–591. [CrossRef]

proceedings

MDPI

Proceeding Paper

Fanning the Flames of Awareness: Strengthening Italian Society's Knowledge on Wildfires †

Francesca Riolfi [1], M. Conceição Colaço [2,*], Ana Catarina Sequeira [2], Leticia Oliveira [2], Fabricio Fava [3] and Iryna Skulska [2]

1 Department of Biology (DiBio), University of Padua, 35122 Padova, Italy; francesca.riolfi@studenti.unipd.it
2 CEABN/InBIO, School of Agriculture, University of Lisbon, 1349-017 Lisboa, Portugal; catarinasequeira@isa.ulisboa.pt (A.C.S.); leticiacoelhooliveira@hotmail.com (L.O.); irynaskulska@isa.ulisboa.pt (I.S.)
3 Research Institute in Art, Design and Society, Faculty of Fine Arts, University of Porto, 4049-021 Porto, Portugal
* Correspondence: ccolaco@isa.ulisboa.pt
† Presented at the XVIII International Seminars on Overarching Issues of the European Area, Porto, Portugal, 23–26 May 2024.

Abstract: Since the late 20th century, wildfires have increasingly disrupted ecosystems, particularly in Mediterranean-climate countries like Italy. This trend is linked to agroforestry and the abandonment of grazing activities, which is worsened by climate change. This study examines the knowledge within Italian society, using data from a survey conducted as part of the European FIRE-RES project, with 152 answers, mainly from northern Italy. The research explores traditional practices, identifies knowledge gaps, and proposes a program to enhance community resilience and to mitigate the impact of wildfires through preparedness and adaptation measures. Results show that, despite a strong cultural connection to the use of fire in Italy, there is a significant gap in public knowledge and readiness to address wildfire incidents among the population. This shows the need to develop educational programs to improve community awareness and adaptive measures to mitigate the impact of wildfires.

Keywords: wildfire risk; education and awareness; community preparedness; survey

1. Introduction

In recent years, the Mediterranean region of Europe has faced significant changes in fire regimes, leading to more frequent, severe, and extensive wildfires [1,2]. These wildfires, which cause damage to forests, human lives, and properties, have become increasingly frequent, severe, and extensive. The region's unique climatic and biophysical characteristics make it particularly vulnerable to wildfires, especially during the hot and dry summer months, when vegetation becomes more fire prone [3,4]. International reports from the Intergovernmental Panel on Climate Change (IPCC) [3] and the European Forest Fire Information System (EFFIS) [5] indicate that rising global temperatures and prolonged summer droughts have exacerbated this hazardous situation. In this scenario, Italy, as one of the Mediterranean countries, is particularly sensitive to these events.

The shift in fire patterns is driven not only by climatic factors, but also by socioeconomic dynamics that have evolved over recent decades in the Mediterranean Basin [3–5]. A significant factor is the migration of the Italian population from rural to urban areas, resulting in major land use changes. Abandoned agricultural and grazing lands have been replaced by shrubs and forest, creating landscapes with continuous fuel sources. This fuel continuity, characterized by tall grasses and scrubland, has contributed to the severity of recent wildfire seasons in Italy [6–8].

2. Materials and Methods

To understand public knowledge and the perceptions of wildfires in Europe, an extensive international survey was conducted as part of the FIRE-RES project. This survey was translated into Italian to encourage participation from Italian residents.

Developed using Google Forms, the survey was open from November 2023 to the end of March 2024. It featured a variety of question formats, including multiple-choice questions, tick-box grids, and checkboxes, with an estimated completion time of nine minutes. The survey comprised 24 questions divided into five sections: respondent characteristics, use of fire in daily life, proximity to wildfires, knowledge about wildfires, and the fire cycle. The snowball sampling methodology [9] was employed to disseminate the survey, through which existing contacts identified further potential participants.

The sample purpose was not intended to be representative of the Italian population, but aimed to explore the general knowledge and perceptions of the respondents to support the development of an international educational platform on wildfire-related topics.

Participants were asked to agree to the terms of the study, ensuring that their data remained anonymous in compliance with GDPR rules (EU data protection).

The answers were analyzed using frequency and descriptive statistics (measures of central tendency as means, variances, standard deviation, percentage values, frequency analyses, ANOVA, and chi-square analysis) using Excel and R.

3. Results

A total of 152 answers were collected and analyzed. The general profile of the interviewees revealed a fairly balanced gender distribution, with 57% aged between 18 and 34 years. A significant portion of the respondents (89%) were from Northern Italy. Additionally, 67% of the respondents were employed, and the majority (82%) lived in urban areas or rural–urban interfaces.

The initial results indicated some interest among the Italian population in the topic, though a significant portion (above 80%) lacked adequate knowledge on prevention, self-protection, firefighting, and post-fire restoration measures, feeling ill-prepared to face extreme events.

The survey revealed that the use of fire in cultural festivities remains deep-rooted in the Italian tradition. Events such as Epiphany bonfires, la "Ndocciata d'Agnone", New Year's Eve celebrations, and Easter and Christmas bonfires showcase the enduring connection between fire and festivity in Italy.

In daily life, fire is primarily employed for cooking (49%), personal use, such as smoking and candlelight (22%), and barbecuing (15%).

The study revealed a moderate interest regarding wildfires, with many individuals reporting seeking information over the past year (54%).

Students reported a statistically significant difference (0.000042) regarding the prevalence of slash–burn activities and other uses of rural fire in their localities compared to the other groups. Regarding the interest in fire and wildfires, the employed group showed statistically significant differences (0.000029) compared to the other three groups.

In most of the topics explored, perceptions and experiences were similar across different groups, regardless of their demographic or geographical categories.

4. Discussion

The data provided a comprehensive insight into the perceptions and behaviors of the Italian respondents towards fire use, cultural traditions, and wildfire awareness. Despite a cultural connection to fire, there is a significant gap in public knowledge and readiness concerning wildfires, which coincides with Spano et al.'s study [10]. Additionally, the regional concentration of respondents from Northern Italy may reflect regional cultural practices and awareness levels, suggesting that future studies could benefit from broader geographical representation to capture nationwide trends.

The numerous traditions of using fire in cultural festivities referred to by the respondents underscore the historical and social importance of fire. This cultural attachment to fire may influence public attitudes toward fire use and safety, suggesting the need for culturally sensitive approaches when designing educational campaigns about fire safety and wildfire prevention. Additionally, the findings indicate that fire remains an integral part of many Italians' daily routines. The reliance on fire for cooking and personal use suggests potential areas for targeted campaigns and interventions to promote fire safety behavior practices and to reduce fire-related risk practices when using fire for these activities outside.

There is a clear need for targeted educational campaigns to improve knowledge about wildfires and the fire cycle, aligning with the results of the Ivčević et al. study [11]. These campaigns should be widely disseminated and tailored to different demographic groups to maximize reach and impact. Campaigns should also take into consideration where the target audience lives, in such a way that the language and fire risk behaviors are adjusted to the public's reality [12]. Initiatives could include community-based programs and workplace training sessions to enhance accessibility and relevance.

The existing literature reinforces these findings, advocating for educational programs that cater not only to different age groups, but also to local community needs [13,14]. Addressing these gaps in future fire education materials is essential for better public preparedness and awareness [15].

5. Conclusions

This study underscores the urgent need for improved wildfire education and preparedness in Italy. Despite a strong cultural connection to the use of fire, there is a significant gap in public knowledge and readiness to address wildfire incidents, also referred to by the respondents.

Socioeconomic factors, such as rural depopulation and changes in land use, increase the vulnerability of certain regions and elderly people to wildfires. Combined with climate changes, these factors have led to more frequent and severe fire events. Additionally, the fragmentation of traditional fire-related practices and knowledge poses a challenge to effective wildfire management.

To address these issues, the next steps should include the development and implementation of educational initiatives, such as FIRE-RES' proposed Fire Education Platform. This platform will serve as a repository for best practices, resources, and data to support ongoing education and community awareness efforts.

Future research should focus on longitudinal assessments of the effectiveness of these educational interventions. Moreover, integrating traditional knowledge with modern wildfire management practices can foster a comprehensive approach to wildfire risk reduction, enhancing community resilience and preparedness.

Author Contributions: Conceptualization, M.C.C., A.C.S., I.S., F.F. and L.O.; methodology, M.C.C., A.C.S. and I.S.; Google form, M.C.C., A.C.S., L.O., F.F. and I.S.; validation, F.R., M.C.C. and I.S.; formal analysis, F.R., M.C.C. and I.S.; data curation, F.R., M.C.C. and I.S.; writing—original draft preparation, F.R.; writing—review and editing, F.R., M.C.C., A.C.S., L.O., F.F. and I.S.; visualization,

F.R.; supervision, I.S. and M.C.C.; project administration, M.C.C.; funding acquisition, M.C.C. and I.S. All authors have read and agreed to the published version of the manuscript.

Funding: The research was funded by the H2020-LC-GD-2020-3-101037419 Project FIRE-RES.

Institutional Review Board Statement: The study followed the Declaration of Helsinki.

Informed Consent Statement: Informed consent was obtained from all participants of the study.

Data Availability Statement: The original contributions presented in the study can be inquired directly to the corresponding author.

Acknowledgments: All authors thank the participants who helped conduct this study. A.C.S. was funded by FCT Postdoc grant FCT/MCTES UIDB/50027/2020. M.C.C. was funded by FCT research contract CEECIND/01072/2020. L.O. was funded by Project FIRE-RES grant H2020-LC-GD-2020-3101037419. I.S. was funded by FCT under research contract 10/2021/UIDP/50027/2020.

Conflicts of Interest: The authors declare no conflicts of interest.

References

1. Ferrara, C.; Salvati, L.; Corona, P.; Romano, R.; Marchi, M. The background context matters: Local-scale socioeconomic conditions and the spatial distribution of wildfires in Italy. *Sci. Total Environ.* **2019**, *654*, 43–52. [CrossRef]
2. Pausas, J.G.; Fernández-Muñoz, S. Fire regime changes in the Western Mediterranean Basin: From fuel-limited to drought-driven fire regime. *Clim. Change* **2012**, *110*, 215–226. [CrossRef]
3. Fernandes, P. Fire-smart management of forest landscapes in the Mediterranean basin under global change. *Landsc. Urban. Plan.* **2013**, *110*, 175–182. [CrossRef]
4. Lee, H.; Calvin, K.; Dasgupta, D.; Krinner, G.; Mukherji, A.; Thorne, P.; Trisos, C.; Romero, J.; Aldunce, P.; Barrett, K.; et al. *Climate Change 2023: Synthesis Report, Summary for Policymakers. Contribution of WG I, II and III to the Sixth Assessment Report of the IPCC*; IPCC: Geneva, Switzerland, 2023.
5. Camia, A.; Durrant, A.; San-Miguel-Ayanz, J. *Harmonized Classification Scheme of Fire Causes in the EU Adopted for the European Fire Database of EFFIS*; Executive report; Publications Office of the European Union: Geneva, Switzerland, 2013.
6. San-Miguel-Ayanz, J.; Durrant, T.; Boca, R.; Maianti, P.; Liberta', G.; Jacome Felix Oom, D.; Branco, A.; De Rigo, D.; Suarez-Moreno, M.; Ferrari, D.; et al. Advance Report on Forest Fires in Europe, Middle East and North Africa 2022. Available online: https://publications.jrc.ec.europa.eu/repository/handle/JRC133215 (accessed on 9 August 2024).
7. Ascoli, D.; Moris, J.V.; Marchetti, M.; Sallustio, L. Land use change towards forests and wooded land correlates with large and frequent wildfires in Italy. *Ann. Silvic. Res.* **2021**, *46*, 177–188.
8. Moreira, F.; Ascoli, D.; Safford, H.; A Adams, M.; Moreno, J.M.; Pereira, J.C.; Catry, F.X.; Armesto, J.; Bond, W.J.; E González, M.; et al. Wildfire management in Mediterranean-type regions: Paradigm change needed. *Environ. Res. Lett.* **2020**, *15*, 011001. [CrossRef]
9. Biernacki, P.; Waldorf, D. Snowball Sampling—Problems and Techniques of Chain Referral Sampling. *Sociol. Methods Res.* **1981**, *10*, 141–163. [CrossRef]
10. Spano, G.; Elia, M.; Cappelluti, O.; Colangelo, G.; Giannico, V.; D'este, M.; Lafortezza, R.; Sanesi, G. Is Experience the Best Teacher? Knowledge, Perceptions, and Awareness of Wildfire Risk. *Int. J. Environ. Res. Public Health* **2021**, *18*, 8385. [CrossRef]
11. Ivčević, A.; Mazurek, H.; Siame, L.; Bertoldo, R.; Statzu, V.; Agharroud, K.; Rego, I.E.; Mukherjee, N.; Bellier, O. Lessons learned about the importance of raising risk awareness in the Mediterranean region (north Morocco and west Sardinia, Italy). *Nat. Hazards Earth Syst. Sci.* **2021**, *21*, 3749–3765. [CrossRef]
12. Colaço, M.C. Reaprender a vivir cos incendios forestais nun contexto de cambio climático. In *Arbores Que Non Arden: As Mulleres Na Prevención de Incendios Forestais*; Proxecto Batefogo—Coord, Ed.; Catro Ventos: Vigo, España, 2019; pp. 127–146.
13. Diekman, S.T.; Stewart, T.A.; The, S.L.; Ballesteros, M.F. A Qualitative Evaluation of Fire Safety Education Programs for Older Adults. *Health Promot Pract.* **2010**, *11*, 216–225. [CrossRef]
14. Lambie, I.; Seymour, F.; Popaduk, T. Young people and caregivers' perceptions of an intervention program for children who deliberately light fires. *Eval. Program Plann.* **2012**, *35*, 445–452. [CrossRef]
15. Pacheco, R.; Skulska, I.; Sequeira, A.C.; Colaço, M.C. Wildfire Education: A Review Across the Globe. In *Fire Hazards: Socio-economic and Regional Issues*; Springer International Publishing: Cham, Switzerland, 2024; pp. 29–41.

proceedings

MDPI

Proceeding Paper

Vegetation Regrowth in Gullies After a Wildfire: The Case Study of the Alva Basin (Centre of Portugal) †

Bruno Martins [1,2,*], Catarina de Almeida Pinheiro [3], Adélia Nunes [1,2], António Bento-Gonçalves [3] and Manuela Laranjeira [3]

1 CEGOT (Centre of Studies on Geography and Spatial Planning), Department of Geography and Tourism, University of Coimbra, 3004-530 Coimbra, Portugal; adeliajnnunes@gmail.com
2 RISCOS (Portuguese Association of Risks, Prevention and Security), University of Coimbra, 3004-530 Coimbra, Portugal
3 Communication and Society Research Centre, Department of Geography, University of Minho, 4710-057 Braga, Portugal; catarina-pinheiro@mail.com (C.d.A.P.); bento@geografia.uminho.pt (A.B.-G.); manuela.laranjeira@geografia.uminho.pt (M.L.)
* Correspondence: bmscmartins@gmail.com

† Presented at the XVIII International Seminars on Overarching Issues of the European Area, Porto, Portugal, 23–26 May 2024.

Abstract: The aim of this study is to identify and characterize gullies considering their morphological and topographical aspects and determine the factors that control vegetation regrowth (VR) in gullies in Alva Basin after the wildfire of 2017. The use of hierarchical clustering identified two groups of gullies. Multiple regression produced three models (R-Square = 81.3%) for gullies group 1, considering the explanatory factors mean width, slope, and burn severity. Group 2 also produced three models (R-Square = 71.8%) but considering the explanatory variables mean width, slope, and flow accumulation. VR mainly depends on post-fire recovery strategies for vegetation, the remaining soil, and site humidity.

Keywords: vegetation regrowth; gullies erosion; sustainability; mediterranean environment

1. Introduction

Gullies contribute very effectively to soil loss and degradation in various areas around the world, e.g., [1]. They are especially effective in certain geomorphic units in arid and semiarid regions, examples of which include the Mediterranean environment. Several studies show that wildfires are responsible for burning extensive areas of scrubland and forests every year, contributing to physical soil erosion, soil degradation, and the presence and formation of gullies. Nevertheless, gullies are sometimes seen as productive hotspots of high biodiversity, playing an important role as an ecological corridor [2]. They are sometimes sediment containment areas, especially when occupied by vegetation. Knowledge of how vegetation recovers along gully channels is therefore particularly interesting, especially in Mediterranean regions that have suffered various forest fires that led to highly degraded soils. This research therefore aims to answer the following objectives (R):

R1: To identify and characterize gullies considering morphological and topographical aspects.

R2: To determine the factors that control vegetation growth, considering a substantial number of geo-environmental factors, including topographic, hydrological, and environmental ones. The stepwise multiple regression (SMR) was used.

https://doi.org/10.3390/proceedings2025113006

2. Materials and Methods

2.1. Study Area

The study area is located in the municipality of Oliveira do Hospital (center of Portugal), more specifically on the right bank of the Alva River. From a lithological point of view, it is essentially composed of granitic rocks with a predominance of coarse-grained porphyroid granite of a calc-alkaline nature and sometimes with megacrystal orientation. The climate of the region is temperate with mild, dry summers and rainy winters. The municipality's area is often affected by forest fires, the most notable in the last decade being the great forest fire of 2017. It started on 15 October and burned an area of 7600 ha of woodland, and it was followed by storm Ana on 10 December 2017, which caused abundant heavy rainfall. According to DGT (2022) [3], the forest is formed mostly by pine trees (52% of the study area) and broadleaf trees (13% of the study area). The pine trees correspond to cultivated forests of Pinus pinaster. In addition, invasive forest species like Acacia dealbata, Ailanthus altissima, occupy 7% of the study area.

2.2. Data Collection

A digital elevation model (DEM) of the study area was created based on contour lines with an equidistance of 10 m. Then, the following topographic parameters were determined: altitude (A), slope (S), aspect (As), topographic ruggedness index (TRI), curvature (C), and flow accumulation (FA). The Normalized Difference Vegetation Index (NDVI) was derived to investigate the post-fire regrowth dynamics of vegetation. Burn severity (BSaf) was also computed. Before performing model processing, the variance inflation factor (VIF) was used for multicollinearity tests, which avoids model variables with the high correlation being simultaneously substituted into the model processing. Landsat 8 OLI/TIRS images with a spatial resolution of 30 m were used to characterize vegetation regeneration (VR) after the fire. Two images were acquired for the pre- and post-fire period, respectively, 28 September 2017 and 30 October 2017. Another five images were included (15 September 2018, 18 September 2019, 4 September 2020, 30 September 2021, and 26 September 2022). The Normalized Difference Vegetation Index (NDVI) was used to analyze VG and the Normalized Burn Ratio (NBR) was used to estimate fire severity (Bsaf). The gullies were identified based on the 2018 orthophotograph provided by the Directorate-General for Territory with a spatial resolution of 25 cm. Overall, 38 gullies were vectorized, and four elements of morphological characterization were calculated, namely total length (L), mean width (mW), maximum width (MaxW), and plan area (Pa).

2.3. Statistical Analysis

Hierarchical clustering (HC) was used to detect homogeneous groups for the 38 gullies identified. Stepwise multiple regression (SMR) was used to identify the determinants of VR in the fourth year after the fire based on 10 independent variables. The TRI variable was removed after applying the VIF. The other variables were integrated and progressively removed.

3. Results

HC identified two groups of gullies based on the five homogeneity parameters Pa, MaxW and mW, S and L. The most important morphological and topographical characteristics of GG1 and GG2 are given in Table 1.

Table 1. Descriptive morphological characteristics of GG1 and GG2.

	Pa (m^2)	MaxW (m)	mW (m)	S (^{0})	L (m)
GG1	1597.0	10.6	5.9	21.6	81
GG2	663.0	7.8	4.0	19.6	146.0

The vegetation in the gullies takes 4 years to reach NDVI values close to those recorded before the 2017 fire, although they are slightly lower (especially in GG2) (Figure 1).

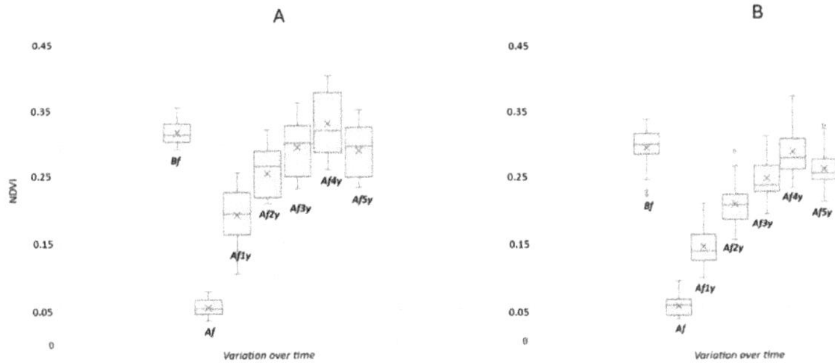

Figure 1. NDVI variation for GG1 (**A**) and GG2 (**B**) (adapt. [4]).

Considering the 11 variables, SMR produced a solution with three models to explain NDVI Af4y. Regarding GG1, the multiple regression produced three models with an R-square of 81.3%, considering the variables mW, S, and BSaf (Table 2).

Table 2. SMR for overall GG1 considering NDVI Af4y.

Model	Variables	Partial R-Square	Model R-Square	Unstandard	Standardized	t	*p*
1	mW	0.584	0.584	0.145	0.546	3.915	<0.001
2	mW	0.195	0.779	0.087	0.753	6.575	<0.001
	S			−0.061	−0.543	−4.867	<0.001
3	mW	0.034	0.813	0.081	0.705	2.705	<0.001
	S			−0.049	−0.439	−3.847	<0.001
	BSaf			−0.041	−0.252	2.321	0.026

Note: The following covariates were considered but not included: L, MaxW, Pa, Rd, A, As, FA, TRI and C. mW is mean width, S is slope and BSaf is burn severity after the fire.

Considering GG2, the multiple regression produced three models with an R-Square of 71.8% considering the variables mW, S and FA (Table 3).

Table 3. SMR for overall GG2 considering NDVI Af4y.

Model	Variables	Partial R-Square	Model R-Square	Unstandard	Standardized	t	*p*
1	mW	0.558	0.558	0.221	0.701	7.347	< 0.01
2	mW	0.117	0.675	0.042	0.685	7.397	< 0.01
	S			0.030	−0.203	−2.189	0.033
3	mW	0.043	0.718	0.400	0.400	3.945	< 0.01
	S			0.313	0.313	2.915	0.005
	FA			0.266	0.266	2.611	0.012

Note: The following covariates were considered but not included: L, MaxW, Pa, Rd, A, As, TRI, C and BSaf. mW is mean width, S is slope and FA is flow accumulation.

4. Discussion

Wildfires have a serious impact on the physical, chemical, and biological properties of soils, which is mainly because the soil is heated and secondarily because of the burning of plants. In the gullies, the NDVI takes 4 years to reach values of vegetation cover close to those recorded before the 2017 fire albeit they are slightly lower (especially in GG2). Other studies of post-fire VR estimated from spectral data in Mediterranean pine forests have recorded similar results. After a disturbance such as a wildfire, herbaceous, perennial, or evergreen plants are the first to become established [2]. This should explain the significant increase in NDVI in the first two years after the fire regardless of BSaf. The increase in NDVI from the third year after the wildfire may be associated with the expansion of Erica and Ulex, species that are particularly resistant to wildfires, with new shoots emerging from the rhizomes shortly afterwards. Overall, after 4 years, the highest VR values occur in the widest gullies (mW), which were distributed over lower altitudes (A) and with higher AF values. The area and greater water availability in these gullies are the main drivers of faster vegetation regrowth. This short-term VR will help to reduce soil erodibility [5] and increase the soil's organic carbon content, thereby improving aggregate stability [6]. Considering VR by gully group, it can be seen that in addition to area, S and BSaf are also important driving forces in GG1, which is made up of the largest gullies in the study area. The results indicate a significant negative relationship between VR and both S and BSaf, indicating that vegetation grows more in areas that are less steep and less affected by fire. In the case of the smallest gully group, GG2, in addition to mW, factors such as S and FA play an important role in VR. This suggests that VR occurs more quickly in gullies on steeper slopes, which favor a greater convergence of water into the gullies. These results once again reinforce the importance of greater water availability in VR.

In our work, although the results indicate NDVI values that are similar after four years to those recorded pre-fire, this does not mean that there have been no qualitative changes in the vegetation and soil.

5. Conclusions

The use of hierarchical clustering (HC) identified two groups of gullies. The first group (GG1) includes gullies with higher values of PA, MaxW, mW and S. The second group (GG2) includes gullies with higher values of L. The vegetation in the gully channel recovered mainly in the two years after the wildfire. In the following years, growth was at a slower rate until it reached similar values in 2021, which was four years after the wildfire that affected the entire study area, eliminating all vegetation. With regard to GG1, multiple regression produced three models with an R-Square of 81.3%, considering the explanatory factors mW, S, and BSaf. For GG2, multiple regression produced three models with an R-square of 71.8%, considering the explanatory variables mw, S, and FA. The results of the study are in line with the evidence that post-fire VR mainly depends on post-fire recovery strategies for vegetation, the remaining soil, and water availability.

Author Contributions: B.M.: Conceptualization, Methodology, Data curation, Writing—original draft preparation, investigation C.d.A.P.: Conceptualization, Methodology, Data curation, Writing—original draft preparation, Investigation. A.N.: Supervision, Methodology, Reviewing and Editing. A.B.-G.: Supervision, Methodology, Reviewing and Editing. M.L.: Supervision, Methodology, Reviewing and Editing. All authors have read and agreed to the published version of the manuscript.

Funding: Centre of Studies in Geography and Spatial Planning (CEGOT), funded by national funds through the Foundation for Science and Technology (FCT) under the reference UIDB/04084/2020.

Institutional Review Board Statement: Not applicable.

Informed Consent Statement: Not applicable.

Data Availability Statement: The data presented in this study are available on request from the corresponding author.

Conflicts of Interest: The authors declare no conflict of interest.

References

1. Montanarella, L.; Panagos, P. The relevance of sustainable soil management within the European Green Deal. *Land Use Policy* **2021**, *100*, 104950. [CrossRef]
2. Romero-Díaz, A.; Ruíz-Sinoga, J.D.; Belmonte-Serrato, F. Physical-chemical and mineralogical properties of parent materials and their relationship with the morphology of badlands. *Geomorphology* **2020**, *354*, 107047. [CrossRef]
3. DGT. *Especificações Técnicas da Carta de Uso e Ocupação do Solo (COS) de Portugal Continental para 1995, 2007, 2010, 2015 e 2018*; Relatório Técnico; Direção-Geral do Território: Lisboa, Portugal, 2022.
4. Martins, B.; Pinheiro, C.; Nunes, A.; Bento-Gonçalves, A.; Laranjeira, M. Site-scale drivers of post-fire vegetation regrowth in gullies: A case study in Mediterranean Europe. *Earth Surf. Process. Landf.* **2024**, *49*, 4371–4387. [CrossRef]
5. Chen, H.; Zhang, X.P.; Abla, M.; Lü, D.; Yan, R.; Ren, Q.F.; Ren, Z.Y.; Yang, Y.H.; Zhao, W.H.; Lin, P.F.; et al. Effects of vegetation and rainfall types on surface runoff and soil erosion on steep slopes on the Loess Plateau, China. *Catena* **2018**, *170*, 141–149. [CrossRef]
6. Guo, M.M.; Wang, W.L.; Kang, H.L.; Yang, B.; Li, J.M. Changes in soil properties and resistance to concentrated flow across a 25-year passive restoration chronosequence of grasslands on the Chinese Loess Plateau. *Restor. Ecol.* **2020**, *28*, 104–114. [CrossRef]

proceedings

MDPI

Proceeding Paper

Symphony of Silence: The Impacts of the COVID-19 Pandemic on Philharmonic Bands in Portugal †

Diogo M. Pinto [1,*], João P. Barreiros [2], André Samora-Arvela [3] and Helena Pina [1]

1 CEGOT, Faculty of Arts and Humanities, University of Porto, 4150-564 Porto, Portugal; mpina@letras.up.pt
2 Faculty of Arts and Humanities, University of Porto, 4150-564 Porto, Portugal; jpbarreiros@letras.up.pt
3 ISTAR, Lisbon University Institute (ISCTE-IUL), Av. Forças Armadas, 1649-026 Lisbon, Portugal; andre.arvela@iscte-iul.pt
* Correspondence: dspinto@letras.up.pt

† Presented at the XVIII International Seminars on Overarching Issues of the European Area, Porto, Portugal, 23–26 May 2024.

Abstract: Philharmonic bands constitute a vital element of Portuguese culture, playing a central role in cultural dynamism, particularly in the country's peripheral spaces. The central objective of this exploratory study is to analyse the impacts of the COVID-19 pandemic on Portuguese philharmonic bands. To this end, a questionnaire was created and distributed among philharmonic bands in Portugal, and so far, 350 responses have been obtained. The first results reveal enormous challenges for the survival of these associations given the withdrawal of a significant number of members, who are now disengaged and unmotivated, which is reflected in the associations' performance, recruitment and financial stability. The results also highlight the strong need for institutional support and new strategies for the sustainability and continuity of these associations.

Keywords: cultural sector; philharmonic bands; COVID-19; impacts; Portugal

1. Introduction

Philharmonics bands are a vital element of Portugal's cultural landscape, playing a central role in boosting cultural initiatives. Their influence on the communities in which they operate is profound, not only due to the variety of activities they promote and make available to the public but also due to their educational role over the centuries [1–4]. Furthermore, they enrich the social and cultural fabric of localities, fostering interaction and social relations between their members [5–8], and are therefore a crucial element of social cohesion in rural areas [9].

The COVID-19 pandemic has posed unprecedented challenges for diverse sectors of society, including cultural and social institutions such as marching bands. In Portugal, these bands are deeply rooted in the musical, educational, cultural and ethnographic landscape [10–12], contributing significantly to the dynamism of local communities, especially in rural areas [13]. This exploratory and unprecedented study aims to understand the pandemic's fundamental social, technical–musical and economic impacts on Portuguese philharmonic bands. By examining these results, we can identify the key areas that call for support and propose strategies to ensure the sustainability of these cultural institutions, which often represent the only cultural and social pillar of localities in the interior of the country, generating a cultural dynamic in settlements which, without them, lose much of their cultural and recreational dynamism [14].

These associations are an impressive element of organising Portuguese society [14]; with the pandemic, they were forced to stop their activity between March 2020 and the second half of 2021. This total stop in activity was a serious problem from the point of view of the financial sustainability of the associations, as they found themselves unable to participate in traditional pilgrimages, their primary source of income. In this context, from an economic point of view, multiple problems emerged in numerous associations, or pre-existing ones worsened, as in many cases, public support was reduced and insufficient. In addition to this problem, there are others related to sociability given the lack of interest of some musicians in returning after an interruption of almost three years. Thus, many associations also face problems from a human resources point of view, with the shutdown caused by the pandemic allowing them to cut from or "bond" with the institution.

2. Materials and Methods

This exploratory investigation was based on applying a semi-structured online questionnaire aimed at philharmonic bands in Portuguese territory after evaluating two philharmonic bands to improve the survey. The questionnaire addressed three fundamental areas: social impacts (participation, social interactions and the recruitment of new members), musical and artistic impacts (quality of performances, adaptation to new rehearsal and performance formats) and economic and financial impacts (financial stability, dependence on institutional support and financing strategies). The questions were structured to cover distinct roles within associations, such as musicians, leaders and conductors, to capture a broad and diverse view of the effects of the pandemic. The questionnaire was distributed to all bands in Portugal, resulting in 350 responses. The data collection process ensured representation of the country's different regions (Figure 1), in addition to allowing a comprehensive territorial analysis of the effects of the pandemic. The questionnaire, in a mixed approach, included several multiple school questions and open-ended questions, thus allowing data related to the complexity of the problems to be obtained.

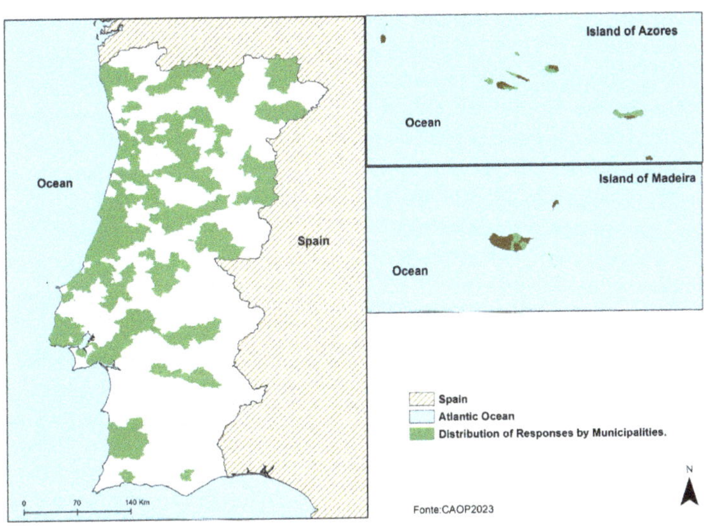

Figure 1. Distribution of responses by municipalities (source: authors).

3. Results and Discussion

Although the questionnaire is comprehensive, only a few results, the most significant ones, are analysed in this article. Thus, the responses correspond to 51.1% (N = 179) from musicians, 40.6% (N = 142) from association directors and 8.3% (N = 29) from conductors.

The geographical distribution covers the mainland and islands of Portugal (Azores and Madeira), exceeding a hundred municipalities.

3.1. The Social Impacts of the COVID-19 Pandemic on Philharmonic Bands

The COVID-19 pandemic has had a significant impact on marching band member participation. Of the respondents, 309 (88.3%) agree that the pandemic negatively affected the participation of musicians. However, 41 (11.7%) totally or partially disagree with this statement. Regarding re-recruiting new musicians, 95 respondents (27.1%) agreed that the pandemic made this process difficult, and 93 (26.6%) partially agreed. Only 49 (14%) disagreed, and 14 (4%) totally disagreed.

Regarding the social interactions between band members, 96 respondents (27.4%) agreed that there were substantial changes, while 85 (24.3%) agreed that the changes were partial. Only 32 (9.1%) disagreed, and 39 (11.1%) totally disagreed. The withdrawal of members during the pandemic was confirmed by 80 respondents (22.9%) who fully agreed and 95 (27.2%) who partially agreed. Only 64 (18.3%) disagreed, and 25 (7.2%) totally disagreed. Thus, it is confirmed that the COVID-19 pandemic profoundly impacted Portuguese philharmonic bands, affecting participation, the recruitment of new musicians, social interactions and the adaptation of new members.

3.2. The Musical and Artistic (Technical) Impacts of the COVID-19 Pandemic on Philharmonic Bands

The pandemic significantly affected the quality of artistic works, particularly in the preservation of performance standards, due to health restrictions, which resulted, for example, in a lack of rehearsals. Consequently, there was a drop in the musicality and cohesion of the groups. In fact, 21.1% (N = 74) of those interviewed positively agreed that their quality was greatly affected. In comparison, 18.9% (N = 66) stated that it was greatly affected, and 5.4% (N = 19) completely agreed that the pandemic affected the artistic quality of their association. In turn, 20.9% (N = 73) reasonably disagreed on this issue, another 20.9% (N = 73) stated that their bandwidth was little affected, and only 12.9% (N = 45) of those interviewed said that the pandemic did not affect their artistic quality in any way. Regarding the mandatory cancellation of events, the results showed that 43.7% (N = 153) of the respondents confirmed that many events were postponed or cancelled due to COVID-19 restrictions, for 28.3% (N = 99), all of them were cancelled, and 14% (N = 49) indicated that some events were cancelled. In comparison, 14% (N = 49) reported that few events were affected. These cancellations have not only harmed the bands' musical performance but also their financial stability, as events are crucial sources of revenue.

The data analysis reveals that the COVID-19 pandemic has substantially impacted marching bands, affecting their artistic qualities. Continuous support strategies that allow them to resume their creative and cultural activities fully should be initiated.

3.3. The Economic and Financial Impacts of the COVID-19 Pandemic on Philharmonic Bands

Concerning the financial sustainability of philharmonic bands, the results demonstrate that despite them suspending activities, the majority were not at risk of closing their activities, at 70.3% (N = 246). However, 13.1% (N = 46) felt this risk. The rest were unable to answer this question, at 16.6% (N = 58). The aim was also to ascertain the financial difficulties these institutions faced due to the pandemic. The results demonstrate that it was difficult for them to pay off their expenses since there were no sources of income for 54.7% (N = 192) of those surveyed. On the other hand, around 32.6% (N = 114) of them considered that the difficulties in paying expenses during this period increased the financial fragility of their bands.

4. Conclusions

The present study demonstrates that Portuguese philharmonic bands faced notable challenges during the COVID-19 pandemic, impacting member participation, recruitment, artistic quality and the financial stability of these associations. Institutional support and the development of new financing mechanisms are crucial to these cultural institutions' survival and continued growth. Recognition of the importance of marching bands is required, as they are cultural pillars and inducers of social inclusion and community dynamism, especially in rural, more peripheral spaces.

To ensure the sustainability of marching bands, it is essential to implement public policies that provide financial, continuous and consistent support to these associations. This includes creating specific financing lines for cultural activities, promoting musical training programmes and encouraging collaboration between bands and other cultural institutions. Population participation also needs to be increased by instituting awareness campaigns, community events and volunteer programmes, strengthening the ties between the bands and the community and supporting each other. Furthermore, the integration of young musicians and the diversification of activities offered by marching bands can attract new members and guarantee the continuity of the musical traditions of this intangible heritage.

Author Contributions: Conceptualisation: D.M.P. Methodology: D.M.P. Formal analysis: D.M.P. Investigation: D.M.P. and J.P.B. Resources: D.M.P. Data curation: D.M.P. Writing—original draft preparation: D.M.P. Writing—review and editing: D.M.P., H.P., A.S.-A. and H.P. Supervision: H.P. All authors have read and agreed to the published version of the manuscript.

Funding: This research received no external funding.

Institutional Review Board Statement: The study followed the Declaration of Helsinki.

Informed Consent Statement: Informed consent was obtained from all of the subjects involved in this study.

Data Availability Statement: The data presented in this study are available at the request of the corresponding author due to privacy restrictions.

Acknowledgments: We thank all of the philharmonic bands and their members who participated in the survey and provided valuable contributions to this study about the impacts of the COVID-19 pandemic.

Conflicts of Interest: The authors declare no conflicts of interest.

References

1. Domingues, A. As bandas filarmónicas na cultura portuguesa. In *Bandas Filarmónicas Portuguesas*; Franco, J.E.D., Ed.; Ancorensis—Cooperativa de Ensino, C.R.L.: Vila Praia de Âncora, Portugal, 2011; p. 11.
2. Franco, J.E.D. *Bandas Filarmónicas Portuguesas*; Ancorensis—Cooperativa de Ensino, C.R.L.: Vila Praia de Âncora, Portugal, 2011.
3. Gomez, A.O. *Contributo das Bandas Filarmónicas Para o Desenvolvimento Pessoal e Comunitário: Um Estudo Efectuado no Alto Tâmega—Sub-Região do Norte de Portugal*; Tese de Doutoramento, Universidade de Vigo: Vigo, Spain, 2007.
4. Madureira, B. Investigação académica sobre filarmónicas e bandas militares em Portugal: Uma panorâmica da situação actual. *Convergências—Rev. Investig. Ensino Artes* **2017**, *10*.
5. Méon, J.-M.; Dubois, V.; Bart, J.-Y. *The Sociology of Wind Bands: Amateur Music Between Cultural Domination and Autonomy*; Ashgate Publishing Limited: Surrey, UK, 2009.
6. Milheiro, M. "Escolas de música, escolas de vida": Estudo de caso na Música Nova de Ílhavo. *Rev. Port. Educ. Artística* **2017**, *7*, 29–44. [CrossRef]
7. Nery, R. *Os Sons da República*; INCM: Lisbon, Portugal, 2010.
8. Pinheiro, C. Importância das Bandas Filarmónicas no Ensino da Música em Portugal: Estudo de Caso (Relatório Final de Mestrado). Instituto Politécnico de Setúbal. 2020. Available online: http://hdl.handle.net/10400.26/36803 (accessed on 3 June 2024).

9. Pinto, D.; Figueiras, C. O contributo das bandas filarmónicas para a construção/preservação da identidade local: O caso de Baião. In *Overarching Issues of the European Space—Preparing the New Decade for Key Issues*; Faculdade de Letras da Universidade do Porto: Porto, Portugal, 2018; pp. 119–138. Available online: https://ler.letras.up.pt/uploads/ficheiros/16754.pdf (accessed on 3 June 2024).
10. Resolução do Conselho de Ministros n.º 56/2013. Diário da República. 2013. Available online: https://data.dre.pt/eli/resolconsmin/56/2013/08/27/p/dre/pt/html (accessed on 4 June 2024).
11. Soares, D.A.B. Importância das Bandas Filarmónicas no Ensino Especializado de Música. Master's Thesis, Escola Superior de Música e Artes do Espetáculo, Escola Superior de Educação, Porto, Portugal, 2020.
12. Sousa, P. *Bandas de Música na História da Música em Portugal*; Fronteira do Caos Editores: Porto, Portugal, 2017.
13. Tavares, C. O Associativismo e a Participação Cívica dos Jovens em Meio Rural. Master's Thesis, Universidade Católica Portuguesa, Viseu, Portugal, 2011.
14. da Silva, E.M.F. Os Filarmónicos: Uma Prática Cultural Amadora [Dissertação de mestrado]. ISCTE—Instituto Universitário de Lisboa. 2021. Available online: https://repositorio.iscte-iul.pt/bitstream/10071/25004/1/master_eva_figueiredo_silva.pdf (accessed on accessed on 4 June 2024).

proceedings

MDPI

Proceeding Paper

Contributions of Territorial and Multilevel Governance: The Case of a Strategic Urban Development Plan in Trás-os-Montes, Bragança (Portugal) †

Maria Patrocínia Correia [1,2,*] and Hermínia Gonçalves [2]

1 School of Public Management, Communication and Tourism, Bragança Polytechnic University, 5300-253 Bragança, Portugal
2 Centre for Transdisciplinary Development Studies (CETRAD), University of Trás-os-Montes and Alto Douro, 5000-801 Vila Real, Portugal
* Correspondence: mferreira@ipb.pt
† Presented at the XVIII International Seminars on Overarching Issues of the European Area, Porto, Portugal, 23–26 May 2024.

Abstract: This paper focuses on one of the instruments provided by the municipal sphere to solve problems in integrated territorial public policies, namely the Strategic Urban Development Plan (SUDP). Considering the starting question, "What lessons can we draw for the construction of governance models in the territorial development processes of historic city centres in low-density cities?", the following objectives were defined: understanding territorial and multilevel governance; capturing the contribution of governance types to the SUDP in Trás-os-Montes, Bragança; analysing the case study through the anchor projects carried out; and understanding the involvement of municipal technical managers in its implementation.

Keywords: governance; territorial; multilevel; cities; historic centre; regeneration; strategic urban development plan

1. Introduction

The plurality of studies on the development of territories reveals their constant transformation based on the economic, social, and political processes that shape and surround them. The territorial scale has become a priority, especially in the political-institutional field, where the surrounding space and geography have emphasised planning at the government level, providing a turning point in public policies. The territorialisation of public policies, namely territory-centred policies, has become a paradigm to be followed. It is based on the need to know and exploit the potential knowledge of a territory and to invest and build institutions through a process involving local actors. Thus, these policies coincide remarkably with those focused on people (people-based policies).

This paper studies an integrated territorial strategy, namely the Strategic Urban Development Plan (SUDP) carried out in the Bragança municipality, the northern interior of Portugal, more specifically, in its historic centre. Once some of the interventions under the SUDP in Bragança were completed, we sought to understand the interventions carried out, reflecting on the governance practises used and trying to respond to the multiple uncertainties that revolve around territorial development. In scientific terms, expanding the understanding of urban development processes in historic centres is urgent. Therefore, we established the territorial and multilevel governance model's influence in Bragança as an object of investigation.

2. Territorial and Multilevel Governance: Concepts and Assessment Dimensions

Territorial governance is defined as a form of "territorial collective action, based on the openness and transparency of the process itself, co-operation/coordination between actors (horizontally and vertically), and on a more or less explicit framework of subsidiarity" [1] (p. 35). This means territorial governance mechanisms must involve not only the community and local organisations but also public and private organisations with the power to influence innovation, such as universities, businesses, and non-governmental organisations. Territorial governance, therefore, depends on the capacity to organise the joint work of these actors in the territory. Thus, the territorial approach to development is required in governance dynamics and understanding the multilevel governance process associated with the territory. The authors warn that multilevel governance must be complemented by "concern about the understanding of territorial arrangements" [2] (p. 51).

There is evidence of three dimensions in territorial governance: the strategic dimension (related to the objectives and strategies of partnerships), the operational dimension (the organisation and management of the structure), and the relational dimension (leadership and participation in the partnership) [3]. Figure 1 presents a set of dimensions summarising territorial governance's analytical components and which were applied to this methodological proposal.

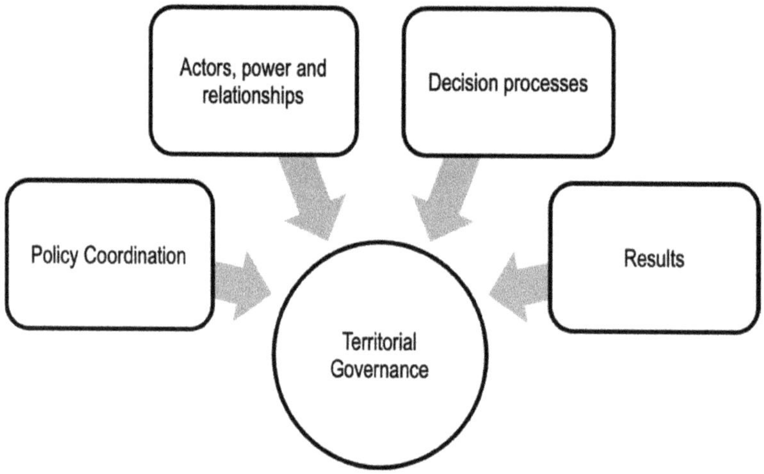

Figure 1. Evaluating the dimensions of territorial governance. Source: [4].

3. Methodological Background and Case Study Context

In the present research, we decided to implement a qualitative methodology in a scientific context. A territorially based instrument was studied, reflected in the SUDP (Table 1). Data collection was conducted in July 2022 and between March 2023 and May 2023, including an analysis of secondary sources extracted from municipality reports, national and regional statistics, and opinion pieces. We decided to select three interviewees based on their proximity criteria to the object of analysis. The interviews were semi-structured and addressed to the Director of the Department of Municipal Services and Construction (DMSC), the Director of the Logistics and Mobility Division (DLM), and the Director of the Construction Division (DC). The three interviews were encoded from I1 to I3.

Table 1. Methodological framework.

Paradigm	Phenomenological and Interpretative Nature
Methods	Qualitative method, case study
Data collection Techniques	Interviews with those who assumed leadership roles in the implementation of the SUDP Documentary analysis of secondary sources, program guidelines, instruments, and reports
Analysis Techniques	Discourse analysis
Case	SUDP (six regional cases)

Own elaboration.

The analysis of the interviews valorised discourse analysis. The subsequent analysis was based on the municipality's technical documents, online news about the works in progress in the city, and the assessment and perspectives of three experts responsible for the design, instruction, execution, and management of the operations/projects.

4. Results and Discussion

According to [3], the management of the development of historic centres can be based on a triangle of efficiency–sustainability–cohesion. This triangle challenges the urban system, emphasising the need to rethink territorial and multilevel governance models, understood as the system of relationships between institutions, organisations, and individuals, ensuring collective choices and their implementation [3]. Multilevel territorial governance will require the creation of structures and, above all, rethinking the relationships between existing structures. Table 2 shows that the discussion of the results considers the dimensions of actors, powers, and relationships; decision processes; and policy coordination [4]. Each of the dimensions has several criteria, and each of the requirements was analysed about its intensity of influence (X—weak to XXX—strong). This table format was chosen for better interpretation.

The authors propose an evaluation matrix on territorial and multilevel governance based on analysing three dimensions: strategic, relational, and operational [3]. The coordination of policies and instruments was decisive for analysing the *strategic dimension*. The intervention of the MB was prioritised in terms of promoting the economic sector, in articulation with urban regeneration projects. It is understood that, as a public entity, the municipality must create conditions for the economic function to settle and develop sustainably. Incorporating a young population in the centre will generate specific needs that must be met to convert it into a set of desirable places to live for an extended period. In this sense, incorporating public spaces, mobility, local commerce, and public services is essential. However, the three interviewees mentioned the lack of private initiative as a negative aspect of the UDP interventions. In the *relational dimension*, the three interviewees, regarding the complementarity and involvement of actors in the development of the SUDP partnerships, believed that this involvement occurred and the partnerships were efficient. At the local level, key public and non-profit organisations, the Higher Education Institution, residents, and retail businesses were involved. Real-time contact platforms were created, receiving communications and alerts from those sharing the spaces and encouraging participation. In the *operational dimension*, the municipality considered that the guiding vision of the SUDP would have to focus on investments that promote economic dynamics and services, adequately complemented by urban regeneration and public space actions such as those associated with mobility.

Table 2. Results of the evaluation of the territorial and multilevel governance of the SUDP.

Dimensions	Criteria	SUDP
Actors, powers, and relationships	Distribution of responsibilities and competences	XXX
	Relation integration	XXX
	Decentralised integration	XXX
	Active participation in civil society	XX
	Flexibility of actors and institutions in the face of change	X
Decision Process	Representativity	XXX
	Democratisation of decisions	X
	Data collective responsibility, with actors contributing to common ends	XXX
	Actions legitimised by cooperation, negotiation, and sharing	XX
	Network governance	XX
Policy Coordination	Decentralisation of policies	XXX
	Horizontal integration	XXX
	Vertical integration	XXX
	Effectiveness and focus on results	XXX
Territorial Governance	Embracing a variety of interests	XXX
	Maximising the effects of policies on society and territories	XXX
	Interaction and collective learning	XX
	Actors as subjects of collective action	XXX
	Focus on improving social cohesion and socio-economic development	XXX
	Territory as a reference framework, with the potentializing of territory	XXX
Multilevel governance outcomes	Horizontal integration (inter-organisational agreements)	XXX
	Horizontal integration (social actors)	XXX
	Vertical integration (intergovernmental agreements)	XX

(X—weak influence to XXX—strong influence), Own elaboration.

5. Conclusions

This article makes several main contributions. Firstly, it is expected that the knowledge from the study will be helpful and fruitful for municipal actors, serving for the development of other strategic urban development processes. The comparative analysis highlights three common topics: development and territorialization based on local problems with integrated intervention, multi-institutional coordination, and horizontal and vertical communication. Secondly, this work was intended to aggregate the practical and scientific components, strengthening the knowledge of territorial development. The last contribution is related to the conviction that this work has shown that the territories and their municipalities will find development opportunities, whether through municipal projects or public policies reflected in community territorial development instruments. Finally, applying the governance assessment tools to several historic centres of low-density cities covered by the SUDP would be a challenge for the future.

Author Contributions: M.P.C. contributed to the research design and implementation, the results analysis, and the manuscript. M.P.C. wrote the study with input from H.G., M.P.C. and H.G. discussed the results and contributed to the final version of this work. All authors have read and agreed to the published version of the manuscript.

Funding: This work is supported by national funds through the FCT—Portuguese Foundation for Science and Technology under the project UIDB/04011/2020 (https://doi.org/10.54499/UIDB/04011/2020, accessed on 2 June 2024).

Institutional Review Board Statement: Not applicable.

Informed Consent Statement: Informed consent was obtained from all subjects involved in this study.

Data Availability Statement: The data presented in this study are available on request from the corresponding author.

Conflicts of Interest: The authors declare no conflicts of interest and the funders had no role in the design of the study; in the collection, analyses, or interpretation of data; in the writing of the manuscript; or in the decision to publish the results.

References

1. Davoudi, S.; Evans, M.; Governa, F.; Santangelo, M. Territorial governance in the making: Approaches, methodologies, practices. *Boletín Asoc. Geógrafos Españoles* **2008**, *46*, 33–52.
2. Dias, R.C.; Seixas, P.C. Territorialização de Políticas Públicas, Processo ou Abordagem? *Rev. Port. Estud. Reg.* **2020**, *55*, 47–60. [CrossRef]
3. Monteiro, S. Mudanças recentes nos processos de governança territorial em Portugal. *Finisterra* **2009**, *88*, 165–173. [CrossRef]
4. Dallabrida, V.R. Governança Territorial: Do debate teórico à avaliação da sua prática. *Análise Soc.* **2015**, *2*, 304–328.

proceedings

MDPI

Proceeding Paper

Preparing Resilient Communities to Mitigate Risk–Bairro Encosta Luz Case Study (Odivelas, Portugal) †

Raul Oliva [1] and Ana Paula Oliveira [1,2,*]

1 Instituto Superior de Educação e Ciências de Lisboa (ISEC Lisboa), Alameda das Linhas de Torres, 179, 1750-142 Lisboa, Portugal; 20210445@alunos.iseclisboa.pt
2 Centro de Investigação, Desenvolvimento e Inovação em Turismo (CiTUR)–Polo Estoril, Avenida Condes de Barcelona, n.° 808, 2769-510 Estoril, Portugal
* Correspondence: ana.oliveira@iseclisboa.pt
† Presented at the XVIII International Seminars on Overarching Issues of the European Area, Porto, Portugal, 23–26 May 2024.

Abstract: Creating resilient communities globally is critical amid increasing disaster risks. The "Resilient Cities" program (RCP) is pivotal in mitigating these risks. Bairro Encosta Luz was surveyed to assess risk awareness among RCP communities, revealing significant gaps in respondent knowledge. Nearly half lacked preparedness training, compounded by vulnerabilities, such as advanced age and limited educational levels. Addressing these gaps is vital, necessitating policies and tools for risk management and social and structural adaptation within Bairro Encosta Luz.

Keywords: resilient cities; community resilience; vulnerability; risk management; mitigation

1. Introduction

Constantly changing global dynamics, marked by social and economic transformations and the increase in disasters, fueled by climate change [1], present challenges to communities worldwide. As a result, building resilient communities in the face of disaster risks becomes imperative on a global scale. Among the strategies devised to mitigate these risks is the establishment of the "Resilient Cities" program (RCP) [2,3]. By 2023, Portugal boasted 54 cities enrolled in the RCP. However, does mere participation in the RCP prepare the city's community to face these risks? Therefore, the objective of this work was to assess the level of awareness of RCP communities regarding the risks they face. A community (Bairro Encosta Luz) from an RCP city—Odivelas—was selected as a case study.

2. Methodology

2.1. Study Area

The study area, Bairro Encosta da Luz (BEL), is located within the municipality of Odivelas (Figure 1). Odivelas, one of the 18 municipalities comprising Lisbon, the capital of Portugal, is a participant in the United Nations' "Resilient Cities" program [3]. Covering an area of 26.4 km^2, the predominantly urban municipality of Odivelas comprises seven parishes (Figure 1). Odivelas ranks as the 15th most populous municipality on mainland Portugal, with 148,034 inhabitants, 47.1% of whom are male and 52.9% female [4].

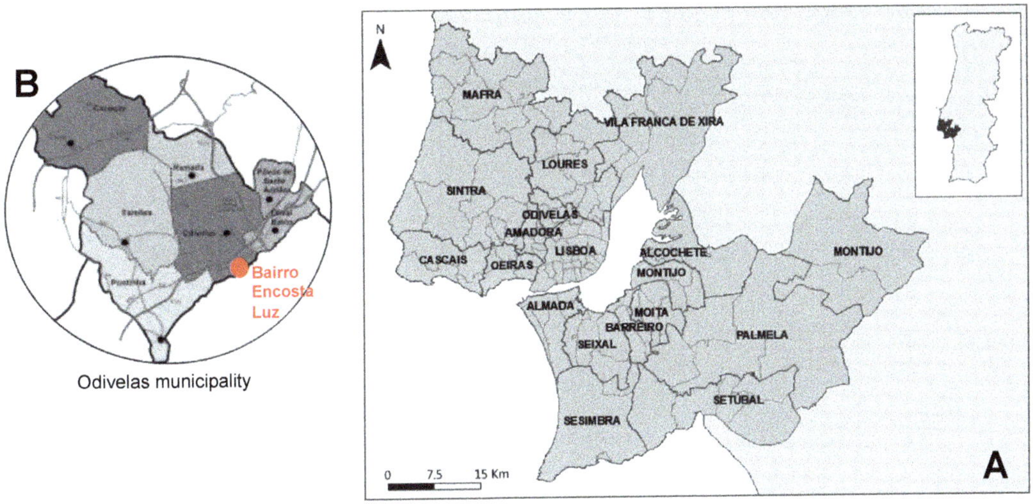

Figure 1. Map of Portugal highlighting the municipalities of Lisbon (**A**) and a detailed view of the municipality of Odivelas with its respective parishes (**B**). Source: adapted from dreamstime.com.

According to the Municipal Emergency and Civil Protection Plan of Odivelas [5], the region is vulnerable to various natural hazards, including floods, wildfires, droughts, intense rainfall, heatwaves, earthquakes, geomorphological incidents, cold snaps, and cyclones/hurricanes. Additionally, anthropogenic risks such as industrial accidents, the transportation of hazardous materials, and nuclear, radiological, biological, or chemical threats, along with domestic accidents like gas leaks, are considered significant. These risks are detailed on the Odivelas Municipal Council website [6], serving as a crucial information resource for the population.

2.2. Data Collection

To evaluate the strategies and measures implemented by the community of BEL in preparing for and responding to disaster risks, a population survey was conducted. The survey, designed using Google Forms®, comprised 32 questions: 25 closed-ended and 7 open-ended, divided into two primary areas: Vulnerabilities and Risk perception.

The survey was carried out in person, utilizing a door-to-door approach, from 19 to 22 September, 2022. A total of 93 responses were collected, representing 1 response per residence and accounting for 60.78% of all residences. The collected data underwent statistical analysis using Microsoft Excel® and the Statistical Package for the Social Sciences® (SPSS) version 28 for Windows. Descriptive statistics, including absolute and relative frequencies, were employed to analyze the survey responses.

3. Results and Discussion

The survey results are summarized in Table 1, with the percentages reflecting the most frequently selected responses. Analysis of these responses allows for several considerations that can be inferred about the demographic profile, residents' perceptions of their geographic environment, the community's assessment of political interventions, and necessary improvements to enhance resilience in their community.

Table 1. Summary of key responses from the survey carried out in the Bairro Encosta da Luz.

	Variable	Bairro Encosta Da luz Responses
Sociodemographic	Sex	Male (55.91%)
	Age	>61 years old (46.27%)
	Education	Basic Education 1st cycle (40.85%)
	Professional situation	Employee (36.56%)
	Type of habitual residence	Villa (72.04%)
	Housing typology	T3 (43.01%)
	Accommodation type	Owner (55.91%)
	Family type	Nuclear-couple and children (32.26%)
	Average number of household residents	2.97
	Residence time	27 to 37 years (29.03%)
Perceptions	Opinion on neighborly relations	Good (54.84%)
	Security actions/courses	No (55.91%)
	Participation in local initiatives	No (90.32%)
	Who to contact in case of emergency	112 (73 responses), Firefighters (22 answers)
	Main risks in the area	Fires, Landslides, Earthquakes
	How do you consider the area in relation to the risks of extreme events	Safe (62.37%)
	Opinion on the municipality's performance	Weak (33.33%)
	Improvement measures	Urban/land cleaning, Legalization of the neighborhood

Analyzing the results presented in Table 1 regarding sociodemographic variables reveals the following trends:

- The population is predominantly male (55.91%) and aged over 61 years (46.27%);
- The majority of the community has completed Basic Education—first cycle (first to fourth year—40.85%), likely attributed to the aging demographic, as many individuals left school early to enter the workforce and support their families, prioritizing livelihood over education;
- A segment of the population remains active and employed (36.56%);
- Villa ownership is prevalent in BEL, with T3 housing being the most common type (43.01%);
- The nuclear family structure, consisting of a couple and their children, is predominant (32.26%);
- BEL has an established community, with residents living in the area for 27 to 37 years.
- Regarding the analysis of the perceptions of the studied population, the following results were obtained (Table 1):
- The BEL population reports positive neighborhood relations, which are crucial for forming resilient communities;
- There is a notable lack of preparation and participation in actions or courses related to risk and safety situations;
- In an emergency, residents would contact 112 and/or the fire department;
- The main risks identified were fire, landslides, and earthquakes; despite these risks, residents perceive their community as safe from extreme events;
- The population views the municipality's performance as weak, contributing to a persistent disconnect between residents and local political authorities;
- To enhance community resilience and preparedness, residents emphasize the importance of cleaning streets and land.

The characterization of BEL as an area of illegal origin adds a critical dimension to the study's context. Beyond the natural risks inherent to its location, the community faces legal and administrative challenges that impact its resilience. The process of legalization could be a key factor in enhancing the community's long-term security and resilience.

4. Conclusions

It is evident that while progress has been made in implementing local risk reduction policies and strategies by "Resilient Cities", there remains significant scope for improvement, particularly in strengthening collaboration between residents and the responsible entities within the territory.

Overall, the BEL community in Odivelas municipality, Lisbon district, demonstrates a limited perception of disaster risks. This is evidenced by inadequate identification of primary risks in the region and a notable gap in preparedness for risk and emergency situations. And, also, in identifying the improvement measures necessary to enhance a more resilient and risk-prepared community: they pinpointed the necessity of street and land maintenance.

Furthermore, there is a prevalent perception of ineffective political performance contributing to a disconnect between local authorities and residents. It is crucial for decision-makers to maintain a continuous commitment to their communities. Likewise, entities responsible for community participation and intervention must refine operational protocols through drills and exercises to bolster response effectiveness during crises. These findings underscore the need for tailored approaches to fortify community resilience and readiness for risk situations, recognizing the unique characteristics and requirements of each community.

It is concluded, therefore, that the studied community faces elevated risk exposure due to factors such as an aging population, population density, outdated urban planning, and an underdeveloped safety culture. Although adherence to the "Resilient Cities" program (MRC2030 program) was seen as beneficial for preparing the municipality and community to cope with risks and extreme events, it is crucial to emphasize the importance of specific and adapted approaches to strengthen community resilience and improve preparation for risk and emergency situations. Enhancing conditions in this community is paramount not only for authorities but also for residents themselves, who bear responsibility for their own safety and that of their neighbors.

Author Contributions: Conceptualization, R.O. and A.P.O.; methodology, R.O. and A.P.O.; validation, R.O. and A.P.O.; formal analysis, R.O.; investigation, R.O. and A.P.O.; resources, R.O.; writing—original draft preparation, R.O. and A.P.O.; writing—review and editing, R.O. and A.P.O.; visualization, R.O. and A.P.O.; supervision, R.O. and A.P.O. All authors have read and agreed to the published version of the manuscript.

Funding: This research received no external funding.

Institutional Review Board Statement: The study was conducted in accordance with the Declaration of Helsinki and approved by the Ethics Committee of Instituto Superior de Educação e Ciências de Lisboa (ISEC Lisboa), Portugal, under code CE/2022/09/01.

Informed Consent Statement: Informed consent was obtained from all subjects involved in the study.

Data Availability Statement: The data presented in this study are available on request from the corresponding author due to the interviewee's privacy.

Acknowledgments: The authors would like to thank the Municipal Council of Odivelas and the Municipal Civil Protection Service of Odivelas. Thanks are also due to Manuel João Ribeiro. The

Bairro Encosta da Luz is part of the international research project "Population Involvement on Risk Management and Mitigation—Testing in Small Communities" (POPIRIM Project).

Conflicts of Interest: The authors declare no conflicts of interest.

References

1. IPCC. *Climate Change 2014: Synthesis Report*; Core Writing Team, Pachauri, R.K., Meyer, L.A., Eds.; Contribution of Working Groups I, II and III to the Fifth Assessment Report of the Intergovernmental Panel on Climate Change; IPCC: Geneva, Switzerland, 2014; 151p.
2. UN. *How To Make Cities More Resilient—A Handbook for Local Government Leaders*; United Nations International Strategy for Disaster Reduction (UNISDR); United Nations: Geneva, Switzerland, 2012. Available online: https://www.unisdr.org/files/26462_handbookfinalonlineversion.pdf (accessed on 15 December 2023).
3. UNDRR. *History of Making Cities Resilient*; United Nations Office for Disaster Risk Reduction (UNDRR): Geneva, Switzerland, 2024. Available online: https://mcr2030.undrr.org/who-we-are/history (accessed on 9 December 2023).
4. INE. *Censos 2021. XVI Recenseamento Geral da População. VI Recenseamento Geral da Habitação: Resultados Definitivos*; INE: Lisboa, Portugal, 2022. Available online: https://www.ine.pt/xurl/pub/65586079 (accessed on 30 November 2023).
5. Resolution of the Council of Ministers No. 39/2015, of August 18. Available online: https://diariodarepublica.pt/dr/detalhe/resolucao/39-2015-70041196 (accessed on 11 December 2023).
6. CMO. *Documentação e Legislação de Odivelas*; Câmara Municipal de Odivelas: Odivelas, Portugal, 2023. Available online: https://www.cm-odivelas.pt/areas-de-intervencao/protecao-civil/documentacao-e-legislacao (accessed on 9 December 2023).

proceedings

Proceeding Paper

The Shared City: Housing and Tourism in the Metropolitan Areas of Lisbon and Porto †

Rui Florentino

CIAUD-UPT, Portucalense University's Branch of the Research Centre for Architecture, Urbanism and Design, 4200-072 Porto, Portugal; ruiflorentino@upt.pt

† Presented at the XVIII International Seminars on Overarching Issues of the European Area, Porto, Portugal, 23–26 May 2024.

Abstract: This document presents the initial developments of a research work on sustainable housing and tourism beyond the urban centers of Lisbon and Porto's metropolitan areas. The preliminary data are taken from the last three Portuguese censuses carried out in the first two decades of the 21st century, in 2001, 2011 and 2021. Maintaining the balance between housing and tourism is a challenge in the governance of these two major cities.

Keywords: city; metropolis; housing; tourism; Lisbon; Porto

1. Introduction

The recent transformation of both Lisbon and Porto's city centers has been subject of significant social debate, mostly between 2010 and 2020—a decade marked by higher touristic pressure [1]. The urban evolution and socioeconomic shift in Porto, as well as in downtown Lisbon, have been the focus of the recent literature on the subject [2]. Nonetheless, the transformations beyond central areas remain unaddressed and need to be considered in terms of sustainable development.

Before the COVID-19 health crisis, there was a huge transformation in these two major Portuguese cities, causing conflicts that affected the urban environment, social and economic activities, housing, citizens and neighborhoods. International tourism has become a key economic element for the metropolitan governance and, possibly, it will become even more relevant in the coming years. This research investigates the case studies of Lisbon and Porto to understand how the balance between housing, tourism and services beyond the central areas is contributing to preserving urban identities and improving sustainable development and governance.

In the present decade, the return of tourism, with visitors attracted by the cultural heritage, has brought challenges to both metropolitan areas. In addition, previous research has shown that some residents have negative perceptions of tourism and the urban landscape [3]. Therefore, it is important to be prepared and manage integrated urban changes through national, regional and local housing policies to ensure a more sustainable method of development, either in Lisbon or in Porto.

The need to balance housing and tourism to solve these kinds of problems links to the idea of shared cities and places that are appropriate for both local inhabitants and tourists. This emerging concept in the field of urban planning represents a contemporary business model, referred to as a shared economy. A larger-scale approach may improve the quality of services in the public realm, for example, through the application of new technologies, which will enable greater access to goods and services, extending the benefits to a wider range of territories and communities.

This paper emerges from an embryo project funded by the research center CIAUD, with a combined methodology using quantitative data and foresight studies, and can be used as an important tool to understand the future impacts of the increase in specialized sectors and its clear impacts on housing for both landowners and tenants. In the framework of the sustainable development goal for cities and communities (SDG nº 11), it is suggested that a shared economy may reduce conflicts and extend its positive effects to a metropolitan area. This goal is connected to the Portuguese competitiveness strategy, especially regarding the exploration of cultural heritage, and also links to regional development strategies: the domain of symbolic capital, technologies and tourism services in the North, and the intention to reinforce the Lisbon brand on a larger regional scale.

2. A General Overview

Through an analysis of the last three Portuguese censuses, it is possible to observe the population growth that has occurred in the two metropolitan areas of Lisbon and Porto in the first decade of the 21st century; however, this growth was more significant in the case of Lisbon (4.9% on the north bank of the Tagus River and 9.1% in the Setúbal Peninsula) than in the metropolitan area of Porto (only 1.7%). Despite the huge variation at the municipal level, in both cases, there was a loss of residents in central cities, almost 10% in Porto and less than 3% in Lisbon, which contrasts with the strong increase in peripheral municipalities; for example, the populations of Cascais and Maia increased by 21 and 12%, respectively.

Between 2001 and 2011, the increase in dwellings was very high and quite homogeneous at the metropolitan level, between 15 and 17%, with the same municipalities mentioned above showing the greatest variation, as well as Matosinhos in the Porto Metropolitan Area. The global financial crisis that occurred at the end of that first decade, particularly in Portugal due to the intervention of the Troika in the forced reduction in sovereign debts, led to an abrupt fall in these indicators in the second decade. In parallel with the large reduction in new housing construction, the phenomenon of global tourism emerged, boosted by "low-cost" travel, which revolutionized the real estate markets in the urban centers of Lisbon and Porto. This had known consequences at various levels, including environmental, social and, naturally, economic factors, with the escalation of prices and the replacement of housing (and inhabitants) with this new type of local accommodation.

In the second decade of this century, the decline in residents was accentuated not only in the central cities of both metropolitan areas but also in Maia, Matosinhos and Oeiras. The local population migrated to the regional outskirts in a centrifugal movement contrasting the invasion of tourists, with residents "passing through" the historic areas of Lisbon and Porto. The increase in the number of dwellings between 2011 and 2021 is residual, with only a 2.2% increase, on average, in the Setúbal Peninsula, a minimal increase in the Porto Metropolitan Area (1.2%) and a miniscule increase, on average, in Greater Lisbon (0.4%). The growth in some municipalities was not enough to satisfy the population's demand, and central cities even experienced a reduction in the total number of houses precisely as a result of the excess of local accommodation.

This phenomenon leads to the concept of a shared city, one that is shared not only between inhabitants and tourists but also by immigrants who come due to the growth of the service economy for foreigners; as such, new urban policies are required. Following on from this introduction, we will discuss the possible spatialization of tourism decentralization in the two metropolitan areas.

3. An Approach to the Cities of Lisbon and Porto

The demand for tourism is centered on historical and heritage areas that present signs of identity and cultural diversity. In the case of the metropolitan area of Lisbon, the heritage is mixed in the central parishes and along the estuary up to Belém and then along the waterfront of Oeiras and Cascais, areas which benefit from an important railway connection enabling daily movements of commuting workers and tourists, such as between the town of Sintra and downtown Lisbon. The central municipality has taken advantage of major international events to promote the capital abroad but is slow to implement a program with an integrated vision to combat the problems of a shared city. Following the consolidation of the new neighborhoods, proposals for further housing construction are debatable and immigration pressure requires urgent measures to ensure the quality of the urban environment.

On a smaller scale, it is possible to understand the potential development of the Alcântara area, which appears to mediate different heritage zones, striking a balance between central density and modern identification. It is an essential area for urban mobility and is able to solve social and housing needs through a concerted program that regulates tourism. The parish has a moderate ratio of local accommodation in relation to housing (5%), as does its neighboring parishes of Ajuda and Campo de Ourique. However, its geographical constraints require special attention to be given to the environmental sustainability of the urban proposals [4].

This ratio of pressure from local accommodation on services like water for housing is also unequal in the cities of Porto and Lisbon. In the five central parishes, the index is above 15%, which prevents new registrations, but in the other ten parishes, there is still a possibility for tourist growth until this "numerus clausus" value is reached. In the general perspective of the Porto case study, the specialization of heritage is also aligned with what happens in the capital region. Given the smaller separation between the banks, the central heritage area extends immediately to the south side of the river but also extends towards the sea and the neighboring municipality of Matosinhos, with peripheral spaces of great cultural interest.

By looking closer, we can see how contemporary heritage emerges in the western part of the area through the architectural works of Pritzker Prize winners and in the urban parks of Serralves and the city, extending to the coast of Matosinhos. The Porto City Council is now developing, in this sense, a strategy that aims to disperse tourist flows through the implementation of six "territorial blocks" in the downtown area and historical center [5]. The so-called "Boavista, Campo Alegre and Marginal" area is located there, situated in the Lordelo do Ouro and Massarelos union of parishes between the center and the mouth, where there are still relevant social challenges. The ongoing projects for urbanization and mobility, if implemented in the articulated way, constitute an opportunity for territorial cohesion between the resident population, services and tourist regulation, utilizing the logic of a shared city.

4. Brief Conclusions

The data initially analyzed have demonstrated the evolution of the number of residents in the main municipalities of metropolitan areas in the last two decades, accompanied by a drastic reduction in the construction of new housing, in contrast to the increase in local accommodation and the massive presence of tourism in the urban center. By disaggregating the indicators at the scale of parishes, the difference is accentuated; for example, in the case of the city of Porto, the number of inhabitants only increases beyond the main circular road in Aldoar, Paranhos and Ramalde.

The "expulsion" of the local population from central parishes is even greater in the case of Lisbon, with losses of more than 20% of residents in Baixa, Misericórdia and Santa Maria Maior, almost 6500 inhabitants, in the last ten years. The dynamics of tourism that cause this substitution is reflected in the real estate distribution, which also has worrying consequences for the wave of immigrants who have arrived to work precariously in services without urban quality in the high-density areas of the Arroios parish.

This problem is being faced in several European cities, with measures and urban policies being implemented, sometimes causing mass disruption, such as the policy recently announced in Barcelona to eliminate local accommodation in a couple of years. In the central municipalities of these two metropolitan areas, there are efforts to limit the number of registrations in the most pressured parishes, but the integration of other planning and mobility actions is necessary in order to avoid the mischaracterization of cultural identity and public space.

At the national government level, it is not yet possible to understand the effects of the "More Housing" and "Build Portugal" programs, which seem to be facing implementation difficulties given the complexity that is inherent in purely financial and fiscal measures without spatialization. But there is no doubt that the procedural simplification of urban legislation can produce results and should be extended to planning and construction regulation.

In terms of this study, Lisbon and Porto are expected to develop new strategic plans, preferably aligned with urban and mobility options, in coordination with the municipalities in their metropolitan area. This is a process that will have to respond, in addition to the usual sectoral studies, to proposals integrated into the economic, social and environmental dimensions of housing and tourism.

Funding: This research was funded by an embryo research project of CIAUD, under the support of FCT, by the references UIDB/04008/2020 and UIDP/04008/2020.

Informed Consent Statement: Not applicable for studies not involving humans.

Data Availability Statement: The data presented in this work were collected in the census inquiries, available on the official page of the Portuguese Institute of Statistics, accessed on the 17 May 2024, www.ine.pt.

Acknowledgments: The author wants to thank architects Fátima Lourenço and Joaquim Flores and student Federico Cuevas, from the Architecture and Urbanism Integrated Master's program at Universidade Portucalense, for their collaboration in the research project.

Conflicts of Interest: The funders had no role in the design of the study; in the collection, analyses or interpretation of data; in the writing of the manuscript; or in the decision to publish the results.

References

1. Tenreiro, J.P. A importância da arquitetura na consolidação dum centro. In *A Baixa do Porto. Arquitetura e Geografia Urbana, 2010–2020*; Book Cover: Porto, Portugal, 2021.
2. Fernandes, J.A.R.; Chamusca, P. *Lisboa e a Airbnb*; Book Cover: Porto, Portugal, 2019.
3. Freitas, I.; Sousa, C.; Ramazanova, M. Historical Landscape Monitoring Through Residents' Perceptions for Tourism: The World Heritage Porto City. *Tour. Plan. Dev.* **2021**, *18*, 294–313. [CrossRef]

4. Florentino, R. O urbanismo à volta de Alcântara-Terra. In *Horta-Navia. Projecto de Intervenção Na Estação Ferroviária de Alcântara-Terra e Zona Envolvente*; Fundação Serra Henriques: Lisboa, Portugal, 2020.
5. Câmara Municipal do Porto. Estratégia de Base para a Dispersão dos Fluxos Turísticos do Destino Porto e a Criação de Quarteirões No Concelho do Porto. 2024. Available online: https://www.cm-porto.pt/files/uploads/cms/Quarteir%C3%B5es%20Tur%C3%ADsticos_Estrat%C3%A9gia%20de%20base%20para%20a%20dispers%C3%A3o%20dos%20fluxos%20tur%C3%ADsticos%20do%20destino%20Porto%20-%20Relat%C3%B3rio%20Global%20-%202024.pdf (accessed on 2 June 2024).

proceedings

MDPI

Proceeding Paper

Environmental Risk Issues at a Coal Waste Spoil Heap—Climate Change Concern †

Paweł Wrona [1,*], Zenon Różański [2], Adam Duda [1] and José Fernandes [3]

1 Department of Safety Engineering, Faculty of Mining, Safety Engineering and Industrial Automation, Silesian University of Technology, 44-100 Gliwice, Poland; adam.duda@polsl.pl
2 Department of Geoengineering and Raw Materials Extraction, Faculty of Mining, Safety Engineering and Industrial Automation, Silesian University of Technology, 44-100 Gliwice, Poland; zenon.rozanski@polsl.pl
3 Department of Geotechnical Engineering, School of Engineering (ISEP), Polytechnic of Porto, Porto and GeoBioTec | UA, 4249-015 Aveiro, Portugal; jap@isep.ipp.pt
* Correspondence: pawel.wrona@polsl.pl
† Presented at the XVIII International Seminars on Overarching Issues of the European Area, Porto, Portugal, 23–26 May 2024.

Abstract: The aim of this study was to identify potential factors related to weather changes and their impact on the coal waste spoil heap. Risk analyses were performed for selected sites using an "if–then" method for the construction of future scenarios, and a risk evaluation matrix for each predicted scenario was determined. The results showed that some of the factors may be ambiguous. For instance, more intense rainfall may cool the material in the heap and reduce the fire risk. On the other hand, it can lead to surface erosion and oxygen access to the material, which may increase the fire risk.

Keywords: spoil heap; climate change; fire risk

1. Introduction

Climate projections for Europe under different Representative Concentration Pathways (RCPs) provide valuable insights into potential future scenarios. Three parameters are crucial temperature, wind velocity, and precipitation. Meinshausen et al. [1] highlight that under RCP2.6, the global mean surface temperature increase by 2100 could be around 1.5 °C, while under RCP8.5, it could reach 4.5 °C relative to pre-industrial levels. Further elaborating on temperature projections, increases ranging from 0.9 to 2.2 °C for RCP4.5 and 1.4 to 2.7 °C for RCP8.5 are indicated to occur by the mid-century, focused around 2055 compared to baseline (1981–2020). The North Atlantic atmospheric circulation patterns, such as the North Atlantic Oscillation, East Atlantic, and Scandinavian patterns, have been identified as influential on the European wind climate at seasonal to inter-annual scales, especially over Western Europe [2]. It has been suggested that the frequency of extratropical cyclones in Northern Europe will align with changes in the storm track regions, leading to alterations in the wind climate [3].

For Poland, according to RCP4.5, the average annual temperature in Poland will increase by 1.3 °C during the current century. According to RCP8.5, the upward trend in average annual temperature is much stronger. Considering projections of precipitation in Poland for the RCP4.5 and 8.5 scenarios, in both cases, the trend is upwards, and rain intensity may be increased.

Coal waste spoil heaps result from mining activity, and they are the subject of wide research, especially due to environmental issues [4]. A spoil heap influences the atmosphere, water, and soil, and the process is expected to be significantly long-term.

Considering future variations in the three crucial meteorological parameters, the influence of climate change on a spoil heap is shown below. An increase in air temperature increases the fire hazard (caused by the spontaneous combustion of coal). Stronger winds make it easier for oxygen to access a spoil heap, which intensifies the self-combustion risk. In addition, stronger winds make dusting and gas emissions more intense. Floods due to runoff from abandoned spoil heaps during heavy rainfall events are expected. Therefore, it is anticipated that this risk will increase due to climate change. Furthermore, it will be a particular issue in those geographical areas where mines are still active, or where closure has occurred recently, so remediation measures such as vegetation, which could mitigate against runoff, are not yet sufficient [5].

However, an analysis of some important factors indicates that their change may have an ambiguous impact on the environmental hazard associated with the heap and crucial examples have been demonstrated in the text.

2. Materials and Methods

Risk identification is the generation of a complete list of risks based on those events or effects that might create, enhance, prevent, degrade, accelerate, or delay the achievement of objectives that are established in the context. In health and safety terms, these events are hazards; in environmental terms, they are impacts [6]. Climate risk management is an approach which looks specifically at climate risks as part of a comprehensive risk management strategy. The risks range from extreme weather events, such as storms and floods, to slow-onset environmental changes such as increasing sea levels and desertification. Risk management in mining areas, especially connected to extreme weather events, allows decision makers to develop specific actions (strategies) that are relevant to safe and economical operating conditions in situations of exposure to a defined hazard, which is often detrimental or harmful [7].

Based on work carried out in the TEXMIN project and a review of the scientific data, we identified climate risk factors (water, temperature, pressure) directly related to climate change and the occurrence of extreme climatic events. The SWIFT technique was used ("IF" and "THEN" formula). Then, the factors were defined with reference to mining coal mining spoil heaps. An example of a selected factor set for a coal waste heap around south Poland is shown in Table 1. It shows various perspectives for a single risk factor.

Table 1. Various perspectives for a single risk factor.

Factor	Direction of Change	Risk	Impact
Rainfall	Increase	IF rainfall increases, THEN increased water erosion may appear at the surface of a spoil heap	Increase in the number of gaps and gullies at the surface leading to more intense oxygen access
Rainfall	Increase	IF rainfall increases, THEN increased water erosion may appear at the surface of a spoil heap	Cools down the burning waste
Temperature	Increase	IF temperature increases, THEN an increased risk of spontaneous combustion is expected	Increase in the number of fire accidents and increased rate of gas emissions

Then, a risk matrix was developed for the years 2020 and 2050 (considering the worst RCP 8.5 scenario). The risk matrix was based on a common combination of probability and consequence [8]: Probability is assigned in the range 1–5, where 1 is rare, 2 is unlikely, 3 is possible, 4 is likely, and 5 is almost certain. The consequence values are also in the range 1–5, where 1 is negligible, 2 is minor, 3 is moderate, 4 is major, and 5 is catastrophic.

Risk is set in the range between I and V, where I is very high, II is high, III is medium, IV is low, and V is irrelevant.

3. Results and Discussion

The example of the analysis and the results are given in Table 2. The selected example due to rainfall shows that a single risk factor should be considered in different aspects. The example due to temperature is clear and one-dimensional.

Table 2. Examples of an ambiguous impact on the environmental hazard—risk evaluated for 2020 and 2050.

Risk Factor	Probability in 2020 (1–5)	Probability in 2050 (1–5)	Consequence	Risk in 2020	Risk in 2050
IF rainfall increases, THEN increased water erosion may appear at the surface of a spoil heap	3	5	3—for increase in gaps and gullies at the surface leading to more intense oxygen access	III	II
IF temperature increases, THEN increased risk of spontaneous combustion is expected	3	5	3—for increase in the number of fire accidents and increased rate of gas emissions	III	II

These results lead to the following observations and conclusions. The variations in risk for the two most important factors, temperature and rainfall, are presented. In both cases, the change in the probability of occurrence in the 2020 and 2050 stalemates was analyzed. The probability in 2020 was determined as possible, and in 2050, as almost certain, in accordance with the predicted RCP scenarios.

Although the factor of rain is crucial and strong water erosion is expected, it can also be expected that very intense rainfall may lead to the partial cooling of the material on the heap (Table 1). Therefore, for this factor, the consequence was assumed to be moderate.

For temperature, the consequence was also determined to be moderate as a change in mean annual temperature of a few degrees Celsius is expected.

However, the most important conclusion resulting from the analysis presented in Table 2 is that in the case of both factors, the risk is expected to increase from III (medium) to II (high).

Author Contributions: Conceptualization, P.W., A.D., Z.R. and J.F.; methodology, P.W., A.D., Z.R. and J.F.; investigation, P.W., A.D., Z.R. and J.F.; writing—original draft preparation, P.W., A.D., Z.R. and J.F.; funding acquisition, P.W. and A.D. All authors have read and agreed to the published version of the manuscript.

Funding: Financial resources from a ministerial grant number 06/030/BK_24/0081 (BK-272/RG3/2024).

Institutional Review Board Statement: Not applicable.

Informed Consent Statement: Not applicable.

Data Availability Statement: Data are contained within the paper.

Conflicts of Interest: The authors declare no conflicts of interest.

References

1. Meinshausen, M.; Smith, S.J.; Calvin, K.; Daniel, J.S.; Kainuma, M.L.T.; Lamarque, J.-F.; Matsumoto, K.; Montzka, S.A.; Raper, S.C.B.; Riahi, K.; et al. The RCP greenhouse gas concentrations and their extensions from 1765 to 2300. *Clim. Chang.* **2011**, *109*, 213–241. [CrossRef]
2. Gonzalez, P.L.M.; Brayshaw, D.J.; Zappa, G. The contribution of North Atlantic atmospheric circulation shifts to future wind speed projections for wind power over Europe. *Clim. Dyn.* **2019**, *53*, 4095–4113. [CrossRef]
3. Moemken, J.; Reyers, M.; Feldmann, H.; Pinto, J.G. Future changes of wind speed and wind energy potentials in Euro-Cordex ensemble simulations. *J. Geophys. Res. Atmos.* **2018**, *123*, 6373–6389. [CrossRef]

4. Różański, Z.; Wrona, P.; Pach, G.; Niewiadomski, A.P.; Markowska, M.; Wrana, A.; Frączek, R.; Balcarczyk, L.; Quintana, G.V.; Ruiz, D.d.P. Influence of water erosion on fire hazards in a coal waste dump—A case study. *Sci. Total Environ.* **2022**, *834*, 155350. [CrossRef] [PubMed]
5. Ribeiro, J.; Flores, D. Occurrence, leaching, and mobility of major and trace elements in a coal mining waste dump. The case of Douro Coalfield, Portugal. *Energy Geosci.* **2021**, *2*, 121–128. [CrossRef]
6. Valverde, G.F.; Duda, A.; Rodríguez, F.J.I.; Frejowski, A.; Todorov, I. Groundwater Risk Assessment in the Context of an Underground Coal Mine Closure and an Economic Evaluation of Proposed Treatments: A Case Study. *Energies* **2021**, *14*, 1671. [CrossRef]
7. Bedford, M.; Foster, P.; Markowska, M.; Kruczek, M.; Zawadzki, P.; Nguyen, P.M.V.; Wrana, A.; Skalny, A.; Janson, E.; Łabaj, P.; et al. *Texmin Handbook: A Guide to Managing the Risks Posed to Working and Abandoned Mining Facilities, and to the Surrounding Environment, Caused by Climate Change*; Główny Instytut Górnictwa: Katowice, Poland, 2022; ISBN 978-83-65503-45-9.
8. *IEC/ISO 31010*; Risk Management—Risk Assessment Techniques. International Electrotechnical Commission/International Organization for Standardization. International Organization for Standardization: Geneva, Switzerland, 2009.

proceedings

MDPI

Proceeding Paper

Innovations in Agri-Food Systems in Europe: Pathways and Challenges to Inclusion and Sustainability †

Iván Tartaruga * and Fernanda Sperotto

Centre of Studies in Geography and Spatial Planning (CEGOT), Faculty of Arts and Humanities, University of Porto (FLUP), 4150-564 Porto, Portugal; fsperotto@letras.up.pt

* Correspondence: itartaruga@letras.up.pt

† Presented at the XVIII International Seminars on Overarching Issues of the European Area, Porto, Portugal, 23–26 May 2024.

Abstract: Optimal functioning of agri-food systems is essential for food security and sustainability. In this sector, Europe faces many issues, such as promoting sustainable and healthy food production in the context of social and economic inequalities. To deal with these issues, we propose a conceptual framework relating to the idea of a regional innovation system considering power relations, called the hierarchical regional innovation system (HRIS). This framework is based on the concepts of eco-innovation, inclusive innovation, and transition as its theoretical foundations. The findings show that the framework can be helpful in European rural contexts.

Keywords: agri-food systems; inclusive innovation; sustainability; hierarchical regional innovation systems; European Union

1. Introduction

Agriculture plays a pivotal role in ensuring food security and environmental sustainability. As the primary source of food production, it holds the key to providing safe, nutritious, and ample sustenance which is necessary for healthy living. However, it also contributes significantly to global greenhouse gas (GHG) emissions, accounting for 26% of the total [1]. In light of this dual role, experts, notably the EAT-Lancet Commission [2], advocate prioritizing healthy diets and sustainable food production as fundamental planetary goals. In this way, many world institutions are striving to resolve this problem through means such as the Sustainable Development Goals—2030 Agenda/UNO, the Paris Climate Agreement/UNO, and the Intergovernmental Panel on Climate Change. Furthermore, some frameworks highlight the relevance of relationships in this field, such as the food–energy–water nexus [3], which shows the synergies and trade-offs of these three elements, as well as their social and environmental effects, and serves as guidance for cross-sectorial development policies.

In the European Union, there are many challenges in the agri-food context. For example, there is urbanisation, globalisation, individualisation of human nutrition, land grabbing, international migration affecting European agriculture, alternative protein production, and so on. However, the main question lies in searching for sustainable and healthy food systems. Failure to do so will result in a global risk to humankind [4]. It is necessary to consider the projected population growth for the upcoming decades, a predicted global burden due to unhealthy and unsustainable food production with impacts on GHG emissions, nitrogen and phosphorus pollution, biodiversity loss, and water and land resources.

Related to the abovementioned issues, Europe faces another challenge because of the megatrends affecting food demand and production, such as emerging consumption patterns, technological change, and competition for natural resources [5]. This context leads to rising social and spatial inequalities in rural areas where many people (mainly farmers) are frustrated with the sustainable and technological policies enforced in their territory, creating large contingents of discontent [6]. Meanwhile, this is in part fed by the production and type of innovations intended for use in the agriculture sector as part of a high-tech and industrialist-centric pattern.

Addressing these imperatives demands a multifaceted approach encompassing social and technological innovations in agricultural practices. Concurrently, recent agricultural protests in Europe against specific environmental sustainability policies underscore the urgency of understanding the underlying causes of discontent within the farming community.

This study provides a conceptual framework for analysing the problems related to environmental and socioeconomic unsustainability in producing innovations in European agri-food systems, showing at the same time the troubles and solutions for different regional contexts in this sector. Drawing upon the theoretical framework of the hierarchical regional innovation system (HRIS) [7], this study endeavours to understand these complexities through the lens of economic geography and innovation studies.

2. Conceptual Framework

Studies on varieties of capitalism have established patterns and relationships between national institutional frames and forms of economic activity [8]. This approach stresses the two main types of varieties: the liberal market economies, found in countries such as the USA or UK, and the coordinated market economies, found in Germany or the Nordic countries. This national-scale framework is a regional version based on the regional innovation system approach [9]. Thus, geographers have created the entrepreneurial regional innovation system and the institutional regional innovation system, i.e., the regional variants of liberal and coordinated market economies.

However, a specific kind of variety of capitalism is relevant to this proposal: the hierarchical market economy variety [10]. Built on Latin American countries, this variety expresses realities in which social and economic inequalities are extreme. Beyond the unevenness, these nations are characterized by diversified business groups within their economies, strong participation by multinational companies, atomistic labour relations, and low educational and professional skill levels.

Based on this variety of capitalism, the HRIS approach was created to consider inequalities and power relations in the innovation process in regions [7]. It uses a concept and a perspective. On the one hand, the inclusive innovation notion defines those innovations that satisfy the needs of poor or marginalised groups [11,12]. In some cases, these groups develop social or technical innovations. On the other hand, the geography of sustainability transitions [13,14] is a territorialised approach to studying transitions. It is a robust improvement over the multi-level perspective [15], a typical approach for transition trajectories of niches, regimes, and macro-trends. Furthermore, this kind of regional innovation system incorporates the idea of eco-innovation [16], i.e., novelties that entail concerns about their environmental impacts and trying to avoid or minimise them.

The HRIS can predominantly highlight two development trajectories underpinned in (ex)inclusion processes (Figure 1). First, it prioritizes inclusivity at its core. Here, inclusion is fundamental to addressing innovations through sustainable and inclusive processes based on more equal terms. The second, in contrast, does not consider inclusion to be central; rather, growth has a strong profit direction. Both trajectories are established in a context of (in)justice and (in)equality defined by power relations and groups engaged in confrontation.

Nevertheless, these settings can lead to some trade-offs or contradictions in themselves. Meanwhile, a more expressive economic dimension (profit-driven) can drive uneven development. On the contrary, a more just process can lead to lower economic growth (because of a non-economic rationale). Thus, this problem can be solved by combining market and embedded relations (economic plus social dimension).

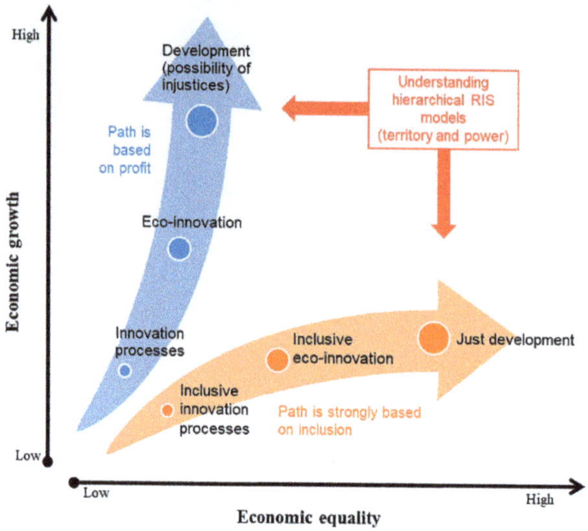

Figure 1. Two development paths that are differentiated by the effectiveness of inclusion [7] (p. 7).

The three main concepts—inclusive innovation, eco-innovation, and territory (HRIS)—are interconnected. First, inclusive innovations underline the differing inclusion levels in the trajectory, from a simple intention to a change in society's mindset (pro-inclusion). Second, the notion of eco-innovation can show, in the context of innovation, niches where entrepreneurs connect resources and business opportunities to sustainable transitions. Third, the HRIS has the role of uncovering relations and place-specific powers that can stimulate or hinder transitions in sustainability.

3. Discussion and Conclusions

The approach presented here can benefit the European agri-food sector. Our findings reveal a deepening rift characterized by economic and technological exclusion, particularly among European family farmers. This exclusionary trend exacerbates rural discontent and has broader societal ramifications, including the proliferation of anti-environmental populism in urban centres [17]. Spatially, rural regions are increasingly experiencing destructuring through processes of territorialization, deterritorialization, and reterritorialization [18].

One major cause of these issues is the focus on technological innovations in Europe, where social and economic inclusion is often overlooked in both private and public sectors, despite community policies such as the New Green Deal and the Smart Specialization platform, which aim to foster a smarter, more inclusive, and greener Europe.

Thus, the HRIS framework is suitable for addressing rural problems. Farmers often remain outside the scope of innovations from multinationals, not only in agriculture but also in commercialization, transportation, and other areas. Additionally, the average farmer has a lower education and skill level compared to urban workers. Therefore, inclusive development processes are fundamental to improving economic competition conditions for these populations, leveraging their specific cognitive assets for social and technical innovations. This study highlights the importance of understanding regional innovation systems, especially regarding inclusion in less innovative regions in both developing and developed countries [19]. Consequently, this understanding can inform and improve policies that leverage regional assets more effectively.

Author Contributions: Conceptualization, I.T. and F.S.; methodology, I.T. and F.S.; validation, I.T. and F.S.; formal analysis, I.T. and F.S.; writing—original draft preparation, I.T. and F.S.; writing—review and editing, I.T. and F.S.; visualization, I.T. and F.S. All authors have read and agreed to the published version of the manuscript.

Funding: This research received support from the Centre of Studies in Geography and Spatial Planning (CEGOT), funded by national funds through the Foundation for Science and Technology (FCT) under the reference UIDB/04084/2020.

Institutional Review Board Statement: Not applicable.

Informed Consent Statement: Not applicable.

Data Availability Statement: No new data were created or analyzed in this study. Data sharing is not applicable to this article.

Conflicts of Interest: The authors declare no conflicts of interest.

References

1. Our World in Data. Available online: https://ourworldindata.org/environmental-impacts-of-food (accessed on 10 March 2024).
2. EAT-Lancet Commission. *Summary Report of the EAT-Lancet Commission*; Stockholm Resilience Centre: Oslo, Norway, 2019.
3. Albrecht, T.R.; Crootof, A.; Scott, C.A. The Water-Energy-Food Nexus: A systematic review of methods for nexus assessment. *Environ. Res. Lett.* **2018**, *13*, 043002. [CrossRef]
4. Willett, W.; Rockström, J.; Loken, B.; Springmann, M.; Lang, T.; Vermeulen, S.; Garnett, T.; Tilman, D.; DeClerck, F.; Wood, A.; et al. Food in the Anthropocene: The EAT–Lancet Commission on healthy diets from sustainable food systems. *Lancet* **2019**, *393*, 447–492. [CrossRef] [PubMed]
5. Kirova, M.; Montanari, F.; Ferreira, I.; Pesce, M.; Diogo Albuquerque, J.; Montfort, C.; Neirynck, R.; Moroni, J. *Research for AGRI Committee—Megatrends in the Agri-Food Sector: Global Overview and Possible Policy Response from an EU Perspective*; European Parliament: Brussels, Belgian, 2019; Available online: https://www.europarl.europa.eu/RegData/etudes/STUD/2019/629205/IPOL_STU(2019)629205_EN.pdf (accessed on 26 February 2020).
6. Rodríguez-Pose, A. The revenge of the places that don't matter (and what to do about it). *Camb. J. Reg. Econ. Soc.* **2018**, *11*, 189–209. [CrossRef]
7. Tartaruga, I.; Sperotto, F.; Carvalho, L. Addressing inclusion, innovation, and sustainability challenges through the lens of economic geography: Introducing the hierarchical regional innovation system. *Geogr. Sustain.* **2024**, *5*, 1–12. [CrossRef]
8. Hall, P.; Soskice, D. (Eds.) *Varieties of Capitalism: The Institutional Foundations of Comparative Advantage*; Oxford University Press: New York, NY, USA, 2001.
9. Asheim, B.; Gertler, M. The Geography of Innovation: Regional innovation systems. In *The Oxford Handbook of Innovation*; Fagerberg, J., Mowery, D.C., Nelson, R.R., Eds.; Oxford University Press: Oxford, UK, 2005; pp. 291–317.
10. Schneider, B. Hierarchical Market Economies and Varieties of Capitalism in Latin America. *J. Lat. Am. Stud.* **2009**, *41*, 553–575. [CrossRef]
11. Heeks, R.; Amalia, M.; Kintu, R.; Shah, N. Inclusive Innovation: Definition, Conceptualisation and Future Research Priorities. *Dev. Inform. Work Pap. Ser.* **2013**, *53*, 1–28. [CrossRef]
12. Heeks, R.; Foster, C.; Nugroho, Y. Introduction: New models of inclusive innovation for development. *Innov. Dev.* **2014**, *4*, 175–185. [CrossRef]
13. Binz, C.; Coenen, L.; Murphy, J.T.; Truffer, B. Geographies of transition—From topical concerns to theoretical engagement: A comment on the transitions research agenda. *Environ. Innov. Soc. Transit.* **2020**, *34*, 1–3. [CrossRef]
14. Truffer, B.; Murphy, J.T.; Raven, R. The geography of sustainability transitions: Contours of an emerging theme. *Environ. Innov. Soc. Transit.* **2015**, *17*, 63–72. [CrossRef]
15. Geels, F.; Schot, J. Typology of sociotechnical transition pathways. *Res. Policy* **2007**, *36*, 399–417. [CrossRef]
16. Kemp, R. Eco-innovation: Definition, measurement and open research. *Econ. Polit.* **2010**, *3*, 397–420.
17. Rodríguez-Pose, A.; Bartalucci, F. The green transition and its potential territorial discontents. *Camb. J. Reg. Econ. Soc.* **2023**, *17*, rsad039. [CrossRef]

18. Haesbaert, R. *O Mito da Desterritorialização: Do "Fim dos Territórios" à Multiterritorialidade*; Rio de Janeiro: Bertrand, Brasil, 2004.
19. Tartaruga, I. Tradition, Inclusive Innovation, and Development in Rural Territories: Exploring the Case of Amiais Village (Portugal). In *Handbook of Research on Cultural Heritage and Its Impact on Territory Innovation and Development*; Oliveira, L., Amaro, A., Melro, A., Eds.; IGI Global: Hershey, PA, USA, 2021; pp. 62–74. [CrossRef]

proceedings

MDPI

Proceeding Paper

Charting the Landscape: Perspectives on Green Innovations in Agriculture †

Fernanda Sperotto * and Iván Tartaruga

Centre of Studies in Geography and Spatial Planning (CEGOT), Faculty of Arts and Humanities, University of Porto (FLUP), 4150-564 Porto, Portugal; itartaruga@letras.up.pt

* Correspondence: fsperotto@letras.up.pt

† Presented at the XVIII International Seminars on Overarching Issues of the European Area, Porto, Portugal, 23–26 May 2024.

Abstract: Concern over the impacts of climate change has significantly increased. While these effects are evident across various contexts, this study focuses on how the scientific literature has addressed the role of green innovations in agriculture. A bibliometric analysis was conducted to understand the structure and evolution of this topic over time. The results confirmed a recent increase in scientific publications and highlighted new research directions, such as identifying the agricultural subsectors most positively impacted by green innovations.

Keywords: eco-innovation; sustainability; scientific mapping

1. Introduction

The effects of climate change are already being felt in all economic activities. In agriculture, the increased incidence of extreme events and the excess and scarcity in water availability have reduced crop expectations and heightened concerns about food security. In addition to hunger affecting about 10% of the world's population, growing urbanisation further reinforces the need to promote changes in the agri-food system [1]. Providing the right technologies to ensure economic, social, and environmental sustainability is crucial for the new agri-food system. The development and dissemination of green innovations aimed at agricultural activities need to contribute greatly to the transformations that the agri-food system needs. These innovations are significant for activities in agriculture because they can mitigate and even nullify the harmful effects of production or consumption on the environment [2,3]. In this context, this study's objective was to analyse how the scientific literature has addressed the role of green innovations in agricultural activities. It is essential to highlight that most studies on green innovations (environmental innovations or eco-innovations) examine industries. For this purpose, a bibliometric analysis was conducted based on articles published on green innovations and agriculture. In addition, we identified the most eco-innovative agricultural activities. This article is organised as follows: Section 2 deals with the methodology applied in the analysis; Section 3 presents the results; and, finally, Section 4 presents the conclusions of the study.

2. Material and Methods

In this study, we examined scientific articles, through a bibliometric analysis, that addressed green innovations in the context of agricultural activities. We collected data from the Scopus database (28 March 2024). The screening included the terms "green innovations" and "agricultural" (including synonyms of these expressions) in the titles, abstracts and

https://doi.org/10.3390/proceedings2025113013

keywords. This resulted in the collection of 569 studies published between 1976 and 2024. We used the Biblioshiny package and Bibliometrix (R software, version 4.4.1.) to measure bibliometric parameters [4]. The research was carried out in two stages: 1st, we elaborated on relational co-citation analyses to detect the theoretical–methodological connections that guide a given research field; 2nd, we performed scientific mapping to observe the conceptual structure behind producing these studies and their thematic evolution according to the research area [5]. To achieve this, we constructed an adjacency matrix and considered two network indices: centrality (which indicates the number of connections that a node establishes in a network) and density (which reveals the proportion of connections). The intersection of these measures resulted in a diagram segmented into four quadrants: the upper-right quadrant represents motor themes (i.e., the most developed and important themes for the research field, characterised by high centrality and density); the lower-right quadrant represents fundamental and transversal themes (which are highly important and linked to other themes in the area but still underdeveloped); the lower-left quadrant represents emerging or declining themes (underdeveloped themes with few connections to other themes, presenting low centrality and density); and the upper-left quadrant represents highly developed but isolated themes [4].

3. Results

We identified 569 studies published between January 1976 and March 2024 that address the theme of green innovations from the perspective of agricultural activities. Overall, scientific research on this theme has grown significantly in the last four years.

Five articles represent the most referenced contributions. The first is a study by Mirabella et al. [6], which analysed the problem of food manufacturing waste and defended the potential of industrial symbiosis to recover waste resulting from food processing. The second study, by Matos and Hall [7], analyses a case study on agricultural biotechnology using the life cycle assessment approach. Padel [8] investigated studies that analysed organic farms in several countries. Notarnicola et al. [9] examined the environmental impacts of food consumption and identified that the agricultural phase is the stage that has the highest impact on all types of food analysed. Finally, Geels et al. [10] discussed the sociotechnical dynamics of low-carbon transitions and highlighted that agriculture activities are part of a very dispersed regime.

Figure 1 presents the evolution of this conceptual field through keyword co-occurrence, subdivided into five periods. Between 1976 and 2010, articles addressing green innovations in agriculture highlighted terms like "innovation", "green innovation", and "sustainable agriculture". From 2011 onwards, new words emerged, such as "sustainable development", "transition", "biofuel", and "food security". Specifically, in the 2015–2018 period, terms like "climate change", "sustainability transition", "circular economy", and "biotechnology" appeared. In the last two segments, it is essential to note the appearance of terms associated with significant actions to mitigate climate change, such as the Sustainable Development Goals and sustainability-oriented innovation.

The synthesis of the thematic map indicates, through the arrangement of keywords, which themes are motor, basic, emerging/declining, and niche themes. As emphasised in Section 2, the intersection of relevance parameters (centrality) and development (density) allows us to identify whether a keyword is related to a more developed field of research (motor themes), is a more transversal term linked to fundamental themes (basic themes), is a term with little development but connected to other research areas (emerging/declining themes), or is a niche theme, i.e., is highly developed but is at a stage of little connection with other themes. Themes guiding studies on green innovations and agricultural activities address aspects related to precision agriculture and the supply chain (Figure 2). In turn, "green innovation", "sustainable development", "climate change", "sustainable", and "sustainable agriculture" were transversal topics linked to fundamental themes. In the quadrant of niche themes, we observed five terms being analysed more in isolation, such as "competitiveness", "absorptive capacity", "corporate social responsibility", and "biofuel". Finally, "absorption" appeared to be underexplored, while "agricultural innovation" seemed to be an expanding term closely related to niche themes.

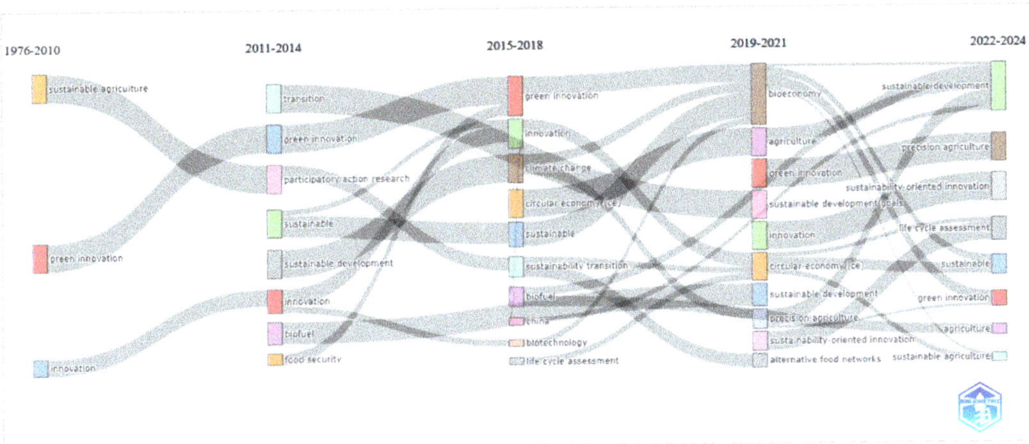

Figure 1. Evolution of the conceptual field of and scientific articles on green innovations and agriculture.

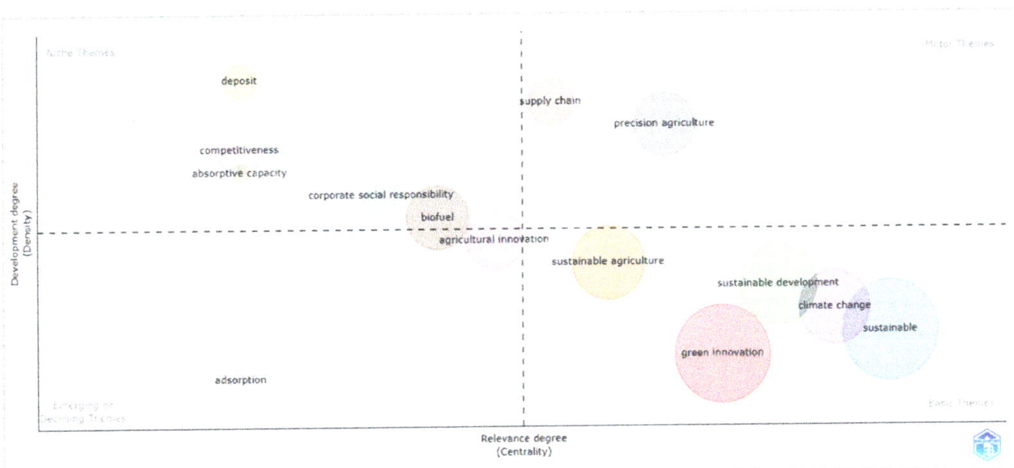

Figure 2. Thematic map according to of scientific articles on green innovations and agriculture.

Beyond scientific mapping, we aimed to identify, through keywords, the activities explored in studies on green innovations and agriculture. In this examination, we identified five categories. The main one encompasses techniques and specific cultivation and exploration segments, such as organic agriculture, organic food, aquaponics, agroecology, bioeconomy, bioenergy, bioplastics, biotechnology, and urban agriculture. The second main category is food production. This category includes aquaculture (salmon, fish farming, and seafood), land cultivation (bananas, cereals, chilli, cocoa, coconut, coffee, legumes in general, lettuce, rice, sugar, sunflower, tea, yeast, soybean, and sorghum), animal farming (honey, goat, and sheep), and industrial processing (meat, wine, and oil palm). The third category relates to renewable energy production, such as biofuels (biogas, biomass, biodiesel, and agrofuel). The fourth group is associated with raw material extraction (bamboo, cotton, and wood) and industrial activities (except food and beverages) such as pulp and furniture. Finally, the fifth group relates to agritourism activities (hotel and rural tourism).

4. Conclusions

This study aimed to analyse how the scientific literature has addressed the role of green innovations in agriculture. The results indicated that, although most articles on the topic are not recent, there has been a rapid growth in the number of publications since 2021. This increased interest in the topic is mainly related to the intensification of climate disturbances and corresponding actions and policies to mitigate their impacts. It was interesting to observe the evolution of the topics analysed in the scientific literature on green innovations and agriculture and to verify their diversity and relevance. Between 1976 and 2010, the topics guiding the studies were mainly focused on just three themes—sustainable agriculture, innovation, and green innovation—but from 2011 onwards, new concerns and perspectives began to guide research, such as a focus on food security, circular economy, and biotechnology. Similarly, the thematic map indicated that research on green innovations and agriculture has focused on topics such as precision agriculture and the supply chain (motor themes), climate change and sustainable agriculture (basic themes), corporate social responsibility and biofuel (niche themes), and agricultural innovation (emerging themes). Another relevant result of this study was the mapping of the activities analysed in the studies, such as specific cultivation and exploration techniques and segments, food production, and renewable energy generation. These subsectors, which are directly or indirectly related to agriculture, are the most prone to and/or demanding of green innovations. In future research, we aim to deepen our analysis by selecting articles that have analysed case studies of green innovations in agriculture.

Author Contributions: Conceptualization, F.S. and I.T.; methodology, F.S. and I.T.; validation, F.S. and I.T.; formal analysis, F.S. and I.T.; writing—original draft preparation, F.S. and I.T.; writing—review and editing, F.S. and I.T.; visualization, F.S. and I.T. All authors have read and agreed to the published version of the manuscript.

Funding: This research received support from the Centre of Studies in Geography and Spatial Planning (CEGOT), funded by national funds through the Foundation for Science and Technology (FCT) under the reference UIDB/04084/2020.

Institutional Review Board Statement: Not applicable.

Informed Consent Statement: Not applicable.

Data Availability Statement: No new data were created or analyzed in this study. Data sharing is not applicable to this article.

Conflicts of Interest: The authors declare no conflicts of interest.

References

1. Food and Agriculture Organization of the United Nations (FAO). *The State of Food Security and Nutrition in the World 2023. Urbanization, Agrifood Systems Transformation and Healthy Diets Across the Rural–Urban Continuum*; FAO: Rome, Italy, 2023. [CrossRef]
2. El Bilali, H. Relation between innovation and sustainability in the agro-food system. *Ital. J. Food Sci.* **2018**, *30*, 200–225. [CrossRef]
3. Tartaruga, I.; Sperotto, F.; Carvalho, L. Addressing inclusion, innovation, and sustainability challenges through the lens of economic geography: Introducing the hierarchical regional innovation system. *Geogr. Sustain.* **2024**, *5*, 1–12. [CrossRef]
4. Aria, M.; Cuccurullo, C. Bibliometrix: An R-tool for comprehensive science mapping analysis. *J. Informetr.* **2017**, *11*, 959–975. [CrossRef]

5. Cobo, M.J.; López-Herrera, A.G.; Herrera-Viedma, E.; Herrera, F. An approach for detecting, quantifying, and visualizing the evolution of a research field: A practical application to the Fuzzy Sets Theory field. *J. Informetr.* **2011**, *5*, 146–166. [CrossRef]
6. Mirabella, N.; Castellani, V.; Sala, S. Current options for the valorization of food manufacturing waste: A review. *J. Clean. Prod.* **2014**, *65*, 28–41. [CrossRef]
7. Matos, S.; Hall, J. Integrating sustainable development in the supply chain: The case of life cycle assessment in oil and gas and agricultural biotechnology. *J. Oper. Manag.* **2007**, *25*, 1083–1102. [CrossRef]
8. Padel, S. Conversion to Organic Farming: A Typical Example of the Diffusion of an Innovation? *Sociol. Rural.* **2001**, *41*, 40–61. [CrossRef]
9. Notarnicola, B.; Tassielli, G.; Renzulli, P.; Castellani, V.; Sala, S. Environmental impacts of food consumption in Europe. *J. Clean. Prod.* **2017**, *140*, 753–765. [CrossRef]
10. Geels, F.; Sovacool, B.; Schwanen, T.; Sorrel, S. The Socio-Technical Dynamics of Low-Carbon Transitions. *Joule* **2017**, *1*, 463–479. [CrossRef]

proceedings

MDPI

Proceeding Paper

Socio-Territorial Fractures and Multi-Scalar Cohesion Policies †

José Manuel Gómez Giménez

Department of Topographic and Cartographic Engineering, Technical School of Engineering in Topography, Geodesy, and Cartography, South Campus, Universidad Politécnica de Madrid, 28031 Madrid, Spain; jm.gomez@upm.es

† Presented at the XVIII International Seminars on Overarching Issues of the European Area, Porto, Portugal, 23–26 May 2024.

Abstract: Current sociodemographic, economic, and ecological dynamics present substantial challenges across Spanish territories. In response, there is a critical need to bolster our capacity for evaluating public policies, enhance the production of actionable data to inform policy design, and reinforce socio-territorial cohesion from a multi-scalar perspective. This approach is pivotal in improving coherence in the pursuit of shared objectives such as equity and sustainability.

Keywords: demographic challenge; development; territorial gaps; geography of discontent; global city; metropolitan area; inequality; residential segregation; affordable housing; mobility

1. Introduction

The recent COVID-19 pandemic has sparked widespread discussion about territorial divides in Spain. Much of the country is grappling with depopulation stemming from outdated economic development models [1] (pp. 58–61). Furthermore, efforts to rejuvenate these regions often rely heavily on commuter mobility, dominated by private vehicles [1] (pp. 78–83). Conversely, escalating inequalities, housing accessibility issues, and residential segregation in major urban areas underscore the repercussions of draining extensive rural territories, thereby undermining broader territorial cohesion [1,2]. These dynamics are emblematic of contemporary economic development, which frequently yields zero-sum effects, underscoring the enduring relevance of dependency theory. Behind these trends lie urbanization processes that have not been effectively planned or guided [3].

Looking ahead, projections indicate mounting challenges. Rural areas are expected to face an old-age dependency ratio exceeding 85% within the next two decades [1] (p. 87), a demographic imbalance that demands correction. Meanwhile, urban areas face bleak prospects in the medium and long term without substantial international immigration. Urgent measures are needed to secure greater stability and socio-economic quality for younger segments of society. This approach represents the most viable and socially acceptable means to enhance work–life balance and align actual fertility rates with desired levels among women. Addressing these intertwined challenges requires an integrated approach, bridging fragmented knowledge and public policies. Understanding their interconnected synergies is crucial for effective governance and planning, minimizing contradictions in public interventions amid constraints on resources, and reinforcing the principle of policy coherence [2].

2. Methods

This paper analyses the repercussions of a tripartite analytical and conceptual model (social, economic, and environmental system), understood as a dystopian ideal type against which future scenarios will be proposed and corroborated using available data for the Spanish case. The consequences derived from these scenarios will entail a series of challenges for which planning and governance should be prepared.

- Social System: Characterized by intersectional stratification, the analysis examines a dystopian scenario marked by escalating inequality, potentially leading to socio-political instability and insecurity. In this context, there would be a significant rise in territorial disparities, residential segregation, and socio-spatial distancing, severely limiting access to opportunities and aspirations for vulnerable groups. Public policies would be challenged to foster equity and socio-territorial cohesion across multiple scales.
- Economic System: Defined by global capitalism, the analysis presents a dystopian model of unconscious developmentalism that fails to justify its role in meeting human needs. This scenario forecasts increasing market tensions and ongoing challenges in housing accessibility, particularly in regions with concentrated job opportunities like agglomeration economies. Amidst growing inefficiencies in meeting basic needs, discontent, uncertainty, and perceptions of relative deprivation would persist. Public policies would face the imperative of demonstrating efficacy in improving overall quality of life.
- Ecological System: Characterized by a planetary biosphere with predominantly finite resources, the analysis highlights a dystopian scenario of ongoing ecological degradation due to inefficient utilization of limited resources and non-circular metabolic processes. Public policies would confront the urgent necessity of promoting a just ecological transition, emphasizing the transformation of unsustainable mobility and transportation models.

3. Results: Challenges and Contradictions

3.1. Socio-Territorial Scenarios and Their Consequences

The demographic transition, facilitated by technological and ideological advancements, has reshaped humanity's population dynamics from high birth and mortality rates to lower birth rates and increased longevity. This transition unfolds in three phases: initially, high birth rates persist due to declining mortality; in the intermediate stages, birth rates gradually decrease while longevity temporarily mitigates losses; and in the final stages, despite low birth rates, losses remain manageable until the exit of baby boom cohorts, often supplemented by significant immigration. In this contextual framework, our welfare provision systems have been designed to be linked to economic and demographic growth. The former could lead to unsustainable long-term outcomes with the current model of consumption and waste, and the latter will ultimately conclude once all societies complete the demographic transition. The challenge here is determined by the importance of long-term socio-historical cycles. Not all societies have experienced development produced by successive ideological and technological revolutions equally. They did not all start at the same time, nor did they begin from the same initial stage or have the same structural starting conditions. Additionally, the speed of induced changes has not been equivalent, nor have the contextual variables or relational structures of interdependence with other actors been comparable [2] (pp. 74–79).

This results in enormous territorial imbalances both between and within countries. In Spain's case, most of its population exhibited all the structural characteristics of the Ancient Regime until the early 20th century. From then on, its development model and population growth in the initial and intermediate phases of the demographic transition led to a consequent concentration of this increase in human capital in major metropolitan areas. For instance, the territory occupied by the six largest metropolitan areas in Spain (Madrid, Barcelona, Valencia, Seville, Málaga, and Bilbao) housed 2.5 million inhabitants in 1900 (15% of the total population), while in 2021, they accommodated 18.2 million people (40% of the total population) [1] (pp. 134–135). The opposite side of this concentrated urbanization is the gradual and continuous depopulation of large territorial areas coinciding with regions of difficult terrain, historical socio-political borders, or areas of significant past extractive value. Today, 85% of Spanish territory faces depopulation dynamics, many in contexts of very low density, and the challenge of structural and migratory over-aging. The overall Spanish population is aging due to the final phase of the demographic transition and the inability of public policies to bring actual fertility rates closer to the desired rates (which, according to CIS surveys, are around the replacement level). This territorial imbalance (Figure 1), along with the differential prospects across various urban regions, is already producing geographies of discontent, translating into different partisan versions of nationalism, regionalism, and localism [2] (pp. 293–438).

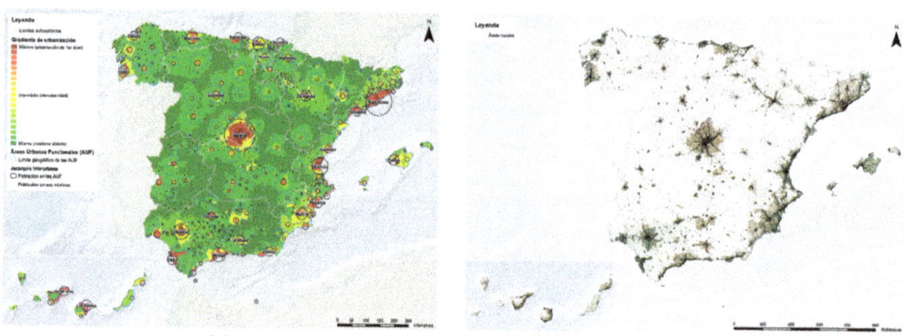

(**a**) (**b**)

Figure 1. Territorial imbalances in Spain: (**a**) Urbanization gradient; (**b**) Urban versus rural.

Major metropolitan areas reproduce models of residential segregation, polycentric decentralization, suburbanization, and increased private vehicle use, exacerbating inequality and socio-spatial distancing [4]. Urban markets' inaccessibility due to rising housing costs forces families to peripheral areas, perpetuating socio-labor conditions that undermine work–life balance, particularly for disadvantaged groups [5]. Immigration, driven by demographic needs, reinforces global labor hierarchies within Spanish urban systems, further widening inequality and spatial disparities. The National Institute of Statistics forecasts that the Spanish population could exceed 54 million in 50 years [6], comprising 40% children of Spanish parents, 20% second-generation immigrants, and 40% international migrants, underlining future demographic complexities. Furthermore, while the richest quintile of the global population holds 83% of the wealth and only has 14% of the young population, the poorest quintile holds only 1% of the wealth but has 25% of the young population. This generational replacement suggests an increase in inequalities and the convergence of these two worlds in global cities. All this foresees an amplification of 'neighborhood effects', meaning a positive feedback loop in the social reproduction of inequality conditioned by the environmental characteristics of primary socialization.

Undoubtedly, this will create socio-political conflicts with unpredictable outcomes. Ignoring these issues will not prevent us from having to face them. Building fully democratic and intercultural citizenship requires civic education in values, social consensus, sufficient resources, and highly innovative political management. Various studies indicate that inequality in Europe has been rising since 1980 [7], approaching levels seen before the two world wars.

3.2. *Tension in Urban Markets: The Housing Problem*

Against this social backdrop, the rise in housing prices and rents in the country's largest cities has accelerated extraordinarily since the COVID-19 pandemic [5]. In rural Spain, over 50% of the housing stock is either vacant or used as holiday homes. In the metropolitan region of Madrid (Figure 2), this percentage represents only 15%, indicating congestion due to incomparable differential demographic growth. The same occurs with the island territories and the Mediterranean arc, further pressured by tourism and the pursuit of profit through domestic savings and international investment. This situation presents a dichotomy: to invest in new models of territorial rebalancing or address the residential emergency in urban areas where the situation is truly urgent. What is clear so far is that market solutions have been ineffective, and social discontent among younger generations and the most vulnerable groups continues to grow.

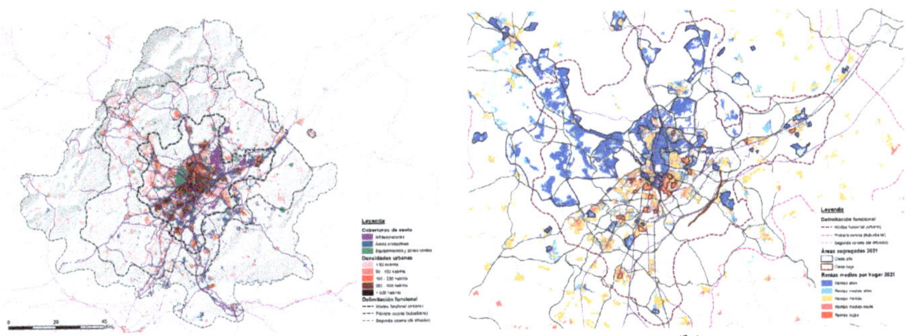

(a) (b)

Figure 2. Metropolitan Area of Madrid: (**a**) Extent; (**b**) Core residential segregation patterns.

3.3. *Transport and Mobility: Technology as an Unsustainable Incrementalist Escape*

Transport emissions in Spain, comprising 30.7% of total emissions, have risen due to increased passenger and freight mobility, briefly interrupted by the 2008 economic crisis and COVID-19 pandemic. This is not merely contextual; in the current climate emergency, CO_2 emissions have once again reached record levels globally.

Since 1950, Spain's urban peripheries have seen rapid population growth, driven by unsustainable car-dependent models. Economic concentration and global pressures have strained urban centers, exacerbating housing crises without effective policies. The proliferation of road infrastructure and private vehicles has further complicated metropolitan mobility, neglected public transport, and increased vehicle use.

The democratization of the automobile and the improvement of road infrastructure, resulting in a network explosion, have led to dynamics of extended urbanization, suburbanization, and changes in the location of economic activity, significantly complicating the needs and patterns of metropolitan mobility. All this has occurred in the absence of territorial planning at the scale of physical transformation processes. Consequently, successive mobility surveys show that private vehicle use continues to rise [4].

Furthermore, we lack sufficient data to evaluate whether the ongoing ecological transition models are as equitable as they should be. It is crucial to understand the demographics of populations that have received public funds for energy efficiency improvements in their homes or for the acquisition of electric vehicles. Despite this, the number of cars in Spain continues to increase, and they are becoming older and more polluting. We cannot burden the costs of the necessary ecological transition on the most vulnerable groups, many of whom were expelled from central areas. This would be unacceptable for a democratic society and could potentially lead to reactionary movements capable of generating significant socio-political instability.

4. Discussion: Complex Context and Opportunity

The Spanish public agenda presents a unique opportunity to analyze the socio-territorial repercussions of economic planning policies within the framework of regional decentralization, which, despite fostering high levels of well-being, has also revealed significant challenges. The restructuring of the interurban system and the decoupling of territorial dynamics from governance have created eco-social issues: the expansion of metropolitan peripheries highlights deficiencies in affordable housing and job stability for younger households, fosters residential segregation, and exposes the lag in adapting public transport to new commuting patterns, leaving ecological transition measures inefficient and generating social discontent among those displaced from central areas. Simultaneously, rural depopulation and the decline of inner urban centers perpetuate extractivist and hyper-productive economic models that, despite emphasizing environmental and heritage value, exacerbate negative externalities and seasonal commuting. Housing shortages and the lack of stable employment in rural areas further compound these issues. To address these crises, the urgent implementation of public policy evaluation methodologies is essential, integrating social, territorial, and ecological dimensions to overcome the current lack of reliable data and avoid ineffective decision-making.

Funding: This communication has been prepared as part of a postdoctoral contract funded by the Next Generation EU funds through the grants for the requalification of the Spanish university system (2021–2023), under the Margarita Salas program for the training of young doctoral researchers at Universidad Politécnica de Madrid (gran number UP2021-035).

Institutional Review Board Statement: Not applicable.

Informed Consent Statement: Not applicable.

Data Availability Statement: The datasets presented in this article are not readily available because the data are part of an ongoing study.

Conflicts of Interest: The author declares no conflict of interest. The funders had no role in the design of the study; in the collection, analyses, or interpretation of data; in the writing of the manuscript; or in the decision to publish the results.

References

1. Gómez Giménez, J.M. *Desigualdades Socioterritoriales en España: La Reconfiguración del Sistema Interurbano Español*; Universidad Politécnica de Madrid: Madrid, Spain, 2023. Available online: https://oa.upm.es/73966/ (accessed on 23 January 2025).
2. Gómez Giménez, J.M. Fracturas Socioespaciales en la Península Ibérica, 1986–2016. Ph.D. Thesis, Universidad Politécnica de Madrid, Madrid, Spain, 22 April 2022. Available online: https://drive.upm.es/s/Db5ct37Fs5Zg0zL (accessed on 23 January 2025). [CrossRef]
3. Gómez Giménez, J.M.; Sá Marques, T.; Hernández Aja, A. Functional urban processes in Iberia: An approach to the integration of the urban territory beyond metropolization processes. *Cuadernos Geográficos* **2020**, *59*, 93–128. [CrossRef]
4. Gómez Giménez, J.M. Urban inequality: Residential segregation in metropolitan contexts without integrated spatial planning. *Cuad. Investig. Urbanística* **2024**, *154*, 1–95. [CrossRef]
5. Martínez, M.A.; Gil, J. Grassroots struggles challenging housing financialization in Spain. *Hous. Stud.* **2024**, *39*, 1516–1536. [CrossRef]

6. National Statistics Institute. Population Projections. Available online: https://www.ine.es/dyngs/Prensa/es/PROP20242074.htm (accessed on 23 January 2025).
7. Piketty, T. *A Brief History of Equality*; Belknap Press: Cambridge, USA, 2021.

proceedings

Proceeding Paper

Landscape, Tourism and Storytelling †

Miguel Castro

CARE—Research Center on Health and Social Sciences, Polytechnic University of Portalegre, 7300-110 Portalegre, Portugal; miguelcastro@ipportalegre.pt

† Presented at the XVIII International Seminars on Overarching Issues of the European Area, Porto, Portugal, 23–26 May 2024.

Abstract: When landscapes tell stories about people, the physical context, and how they become part of the identity of the inhabitants, the land acquires meaning for tourists due to their authenticity. Monument Valley and Alentejo are places where identity and landscape are crucial for touristic activity. It remains to be seen whether the future is massification or whether it will be able to maintain a balance between landscapes, culture, and sustainable tourism. From a Humanistic Geography point of view, we have conducted research based on in-depth interviews that show an unstable balance between a way of life and having economic income.

Keywords: tourism; storytelling; landscape; Monument Valley; Alentejo

1. Introduction

A key aspect of the success of tourism is the landscape and the experiences it provides. The landscape's ability to tell the visitors a story, valuing the populations that frame the destinations, completes the attraction of those regions. Landscapes are not static photographs; they tell the story of people and their relationship with the physical context. When landscapes become identity for populations, they also acquire meaning for tourists due to their authenticity.

We focused on two regions, Monument Valley, USA, and Alentejo, Portugal, where the identity of the populations, the physical context, and their history are responsible for the sustainable development of the tourist activity.

Using a perspective linked to Humanist Geography as a theoretical framework, we developed a set of interviews with privileged actors of these two destinations to try to understand how the tourist phenomenon is accepted in these territories.

2. Brief Theoretical Framework

The concept of landscape appeared by the end of the 15th century and was essentially related to rural, bucolic, and natural places. Nowadays, landscape has a double meaning. On the one hand, we can define it as a portion of the Earth's surface that people contemplate, made up of natural and/or human elements; on the other hand, the concept is more complex; the [1] European Landscape Convention (2000:2) defines it as "part of the territory, as perceived by populations, whose character results from the action and interaction of natural and/or human factors". Quoting [2] Barata Salgueiro, "If the starting point is an objective observation of a portion of the Earth's surface, the need to explain the visual elements transports us to cultural aspects, territorial organization, or other equally significant aspects".

We align our methodology with the Humanist and Culturalist Schools, which look at reality in an individualized way and within a cultural context. These schools state that the concept of landscape is not only a cultural product but also the result of an individual experience, which comes from the senses, the way we feel and connect with the space that surrounds us. The subjective, sensorial, and individual elements make people build their relationship with the area where they live. The landscape influences the behavior and way of life of a population; the territory acquires multiple meanings, like a sense of belonging and a collective and individual (hi)story; the landscape also becomes an identity. In this approach, we highlight [3,4] Yi - Fu-Tuan.

Monument Valley and Alentejo are demonstrations of this symbiosis between space and people, and both are isolated spaces. Our aim here is to attest the following statements:

- The landscapes where the Navajo and Alentejanos built their way of life are closely linked to the construction of the identity of these populations;
- The culture of these people only acquires meaning in a specific territory;
- Isolation causes an intimacy with space that leads to the construction of a world view inseparable from the man/space relationship;
- Tourism is an activity that allows for the maintenance of particular cultures.

3. Methodology

In this investigation, we adopted a qualitative approach based on interviews with privileged resident observers and actors with influence in the territory, in addition to bibliographical research and direct observation. The qualitative research and content analysis allow us to see extracts from conversations and interviews: [5] ((...) "causal explanations (...) of objects, events or situations not necessarily considered representative", making them advantageous because of the fact that "relationships can be established in many and varied situations" According to [6] Krippen-droff (2004), content analysis is an investigation technique that allows for validating information obtained through multiple resources, whether texts (books, newspapers, or interviews) or the meanings that we can extract from the narratives that express the visions of the actors interviewed.

We conducted eight interviews (4 in Monument Valley and 4 in Alentejo) with tour operators, tour guides, wildlife photographers, and prominent members of the community; we also consulted storytelling books by members who are unanimously considered to be guardians of traditions by the council of elders of the Navajo people.

In the interviews and documents, we were privileged to see the messages and reflections related to the concept of topophilia and sense of place, expressed by [3,4] Tuan. Refs. [7,8] Haesbaert (2003, 2004) also states that territoriality is based on the emotional and affective relationship that the individual establishes with the territory on a daily basis; from this relationship, a sense of place is and becomes identity.

In the case of the Navajo people (Diné, as they call themselves), the connection to territory is intensely experienced and very clear in the interviews, as well as in the narrative of John Holiday [9]. One of the memories that persists is the "Long Walk". Between 1863 and 1866, the Diné were forced to leave their territory to Bosque Redondo. The US military coerced men, women, and children to walk 400 to 725 km (depending on the route) from their homeland. Despite the iron fist to which they were subjected, after many revolts, they managed to return. According to them, the Navajo people emerged from the lower world to its current territory (Dinétah), the place on Earth where the people and the "Holy People" interact. It is this relationship that still sustains and bases their culture and way of life.

The interviews show that even though they returned to their homeland, there was a modification of the traditional form of life. This fact led to a disruption of the social pattern and opened up the door for an assimilation of cultural traits of a white and capitalist America, leading to a hybrid cultural path, although with a Diné identity.

In Alentejo, the reality of the traditional way of life and the landscape that shaped it are still present. The region is one of the poorest and isolated of Portugal. For decades, these people were exploited by the Terratenientes, who owned thousands of hectares and forced the workers to put up with hard work, from sunrise to sunset, for little money, under the harsh southern sun. Their diet was poor and based on dry bread, wild herbs, eggs, and some animal fat. The extensive peneplain of Alentejo calls for calm, and the gastronomy, somewhat modified, but not too much, is a hymn to the survival capacity of the people. They also have a unique musical expression (Cante), World Heritage by UNESCO, which reflects the sadness of their landscape and depicts everyday scenes of an experience of the past, but that still fits in its surroundings.

These people have a strong attachment to their origin and way of life, which is embedded in the landscape. They have a feeling of topophilia and sense of belonging, which is something that amazes tourists and is appreciated and valued. Tour operators condense the story of the territory and emphasize the quietness and immensity of the landscape, where everyone has time to calmly enjoy the peace, soft light, wine, and gastronomy, or just to be with the family.

4. Results and Discussion

Geographically, in their specific territorial contexts, the two regions are isolated and forgotten by the policies of regional and central governments. The isolation of these spaces in relation to the national whole is accentuated. The dimension of the USA in relation to the Navajo Nation is more significant than in Alentejo, but both have a landscape, origin, and geostructural dynamics that are different from the rest of the national territories, which makes them unique in their contexts.

The isolation of these spaces led to the genesis of different cultures and ways of life in relation to the rest of the countries. Although the Navajos people have a specific culture and way of life, they have some common traits with other native Americans tribes; however, they are distinctive enough not to share a single cosmogony and beliefs. In fact, the Navajo cosmogony comes from the landscape itself.

From the interviews, literature review, and direct observation, the main ideas that emerge are related to a strong feeling of belonging and topophilia. Ways of life and traditions only acquire meaning in these landscapes' contexts. Those who move away from these territories due to isolation and lack of economic and work opportunities remain linked to their place of origin and are, in a way, ambassadors and promoters of their cultural traits and landscapes.

Diné interviewees assume two positions with points of contact, but with significant differences. The elderly feel that "they were not able to do a good job on passing on the traditions to the younger ones". Young people and adults feel discouraged, unable to be Navajo but not feeling like Americans either. In the reservation, unlike other American citizens, they do not have the same opportunities of getting a job or having access to housing, water, electricity, or education. They feel discriminated again and complain of having been exploited by uranium mining on their lands without having received proper compensation. The government of the Navajo Nation does not fulfill the expectations of the majority of the population, getting lost in bureaucracy that increases distances between Dinè and their representatives. From the central government, through the BIA (Bureau of Indian Affairs), they do not receive answers to the widespread poverty of the population living in their territory. Tourism, while not being a solution to their problems, manages to provide some employment and visibility to these people and their heritage, attracting attention to their reality.

In Alentejo, the situation is different. The territory and population, although with levels of development below the national average, have undergone positive changes that raise the standard and quality of life not comparable to the reality of the past. Tourism has not only been a path to economic growth and improved standards of living but it has also contributed to the maintenance and appreciation of traditions and ways of life. The region has managed to obtain significant gains from tourism and simultaneously preserve its cultural heritage, having been considered Intangible Heritage of Humanity by UNESCO (the Cante Alentejano, Figured in Clay of Estremoz, the production of rattles, and, in another field, the Alqueva Dark Sky Reserve).

In both regions, tourism has been growing and providing increased visibility to these territories. Tour operators have constructed a narrative based on lifestyles, traditions, and landscapes that appeals to the unique and, in some way, exotic nature of the regions within the national whole. If in the Navajo territory the regional government and tourist operators have managed to maintain a relatively balanced tourist flow, with the persistence of the traits that make them unique, in the case of Alentejo, due to the smallness of the national whole, tourist flows have been increasing due to the national and international dissemination of the history of a population still linked to tradition and framed in a calm, peaceful landscape.

5. Conclusions

The culture, history, and way of life of the populations that inhabit the territories in question are a result of the isolation and the landscape, which became identity for them. Outside their territory, all the references disappear due to the absence of geographical contexts.

The Diné population feels somewhat abandoned and with no government support.

In Alentejo, however, there is a general appreciation of its culture and heritage.

Tourist activity could end up being a sword of Damocles. If, on the one hand, it allows for the maintenance of these realities, on the other hand, a lack of moderation of the tourist load could jeopardize these unique people's way of life and their territorial framework.

The main conclusion is that both populations show a deep sense of belonging and a topophilia connection to their territory of origin.

Funding: This research received no external funding.

Informed Consent Statement: Informed consent was obtained from all subjects involved in this study.

Conflicts of Interest: The author declares no conflicts of interest.

References

1. Council of Europe. *European Landscape Convention*; Council of Europe, Cultural Heritage, Landscape and Spatial Planning Division: Strasbourg, France, 2000.
2. Barata Salgueiro, T. Paisagem e Geografia. *Finisterra* **2001**, *72*, 37–53.
3. Tuan, Y. *Topophilia: A study of Environmental Perception, Attitudes and Values*; Prentice-Hall: Englewood, NJ, USA, 1975.
4. Tuan, Y. *Espaço e Lugar: A Perspetiva Da Experiência*; Difel: Sao Paulo, Brasil, 1983.
5. Velez de Castro, F. Constructing Territories of Deterritorialization–Reterritorialization in Clarice Lispector Novels. *Soc. Sci.* **2022**, *11*, 575. [CrossRef]
6. Krippendorff, K. *Content Analysis: An Introduction to His Methodology*; Sage: Thousand Oaks, CA, USA, 2004.
7. Haesbaert, R. *O Mito Da Desterritorialização, Do "Fim Dos Territórios" à Multiterritorialidade*; ACADEMIA: Bertrand, Brasil, 2004.

8. Haesbaert, R. Da desterritorialização à multiterritorialidade. *Bol. Gaúcho De Geogr.* **2003**, *29*, 11–24.
9. Holyday, J.; McPherson, R.S. *A Navajo Legacy: The Life and Teachings of John Holiday*; University of Oklahoma Press: Norman, OK, USA, 2005.

proceedings

Proceeding Paper

The Common Agricultural Policy Strategic Plan Should Be Oriented More to the North: Analysis and Synthesis of the Manifestos of the Farmer Civic Movements in the Northern Region †

Bruno Leitão * and José Vieira

Norte of Portugal Regional Coordination and Development Commission, 4150-564 Porto, Portugal; josevieira@ccdr-n.pt
* Correspondence: bruno.leitao@ccdr-n.pt

† Presented at the XVIII International Seminars on Overarching Issues of the European Area, Porto, Portugal, 23–26 May 2024.

Abstract: Farmers are an important and fundamental professional class on which the food, health and well-being of all citizens depend. Recently, and similar to what has happened in other European countries, in Portugal there were also demonstrations of civic and spontaneous movements of farmers dissatisfied with the current Common Agricultural Policy Strategic Plan (PEPAC), with the excessive bureaucracy and with the increase in the costs of production and unfair competition from agricultural products from other countries, especially from outside the European bloc, which do not meet the same production requirements and which are available to consumers. This work consists of a synthesis and reflection on the reasons, arguments and demands made by civic movements of farmers in the northern region of Portugal (NUT II). Among the most frequent allegations was the need to consider specific aspects of the northern Portuguese region, small family farming and some specific sectors. Although the factor that triggered farmers' indignation was the reduction in support payable under eco-schemes, the consideration of the common land areas for subsidies and for certification in organic farming (in the mountain region), the installation of anti-hail towers in the fruit growing area and compensation for increases in energy and animal feeding costs are among the most frequently mentioned reasons for which farmers require answers.

Keywords: agriculture; strategic plan for the Common Agricultural Policy; north of Portugal; farmer civic movements

1. Introduction

In a period of pandemic (January 2020 and May 2023) and the context of war in Ukraine since February 2022, with an impact on the prices of production factors and agricultural products, the new Common Agricultural Policy (CAP) for 2023–2027, with application starting in January 2023, was seen early on as a reinforcement of fair, sustainable and competitive agriculture that was considered greener.

In October 2023, discontented European farmers started their protests against the new CAP (2023–2027), first in the south of France (Occitania), spreading to the rest of the country and moving on to other Member States such as Poland, Belgium, Germany, Italy, the Netherlands, Romania, Spain and Portugal [1,2]. This discontent and unrest may well be associated with a regional development trap (decline in economic growth, employment and

https://doi.org/10.3390/proceedings2025113016

productivity relative to their neighborhood regions) leading to the increase in Eurosceptic parties [3].

Reflection on the reasons for these protests, particularly those relating to the regional reality of the NUT II north of Portugal, was based on the interpretation of manifestoes and information conveyed by civic movements of farmers and then om analyzing, synthesizing, interpreting and decoding the reasons and origins of the problems that led them to hold demonstrations in February 2024.

Independent of agricultural organizations, the civic movements of farmers brought to the streets hundreds of farmers dissatisfied with the first steps of the Common Agricultural Policy Strategic Plan (PEPAC). This plan began to be implemented at the beginning of 2023 and up to the end of the year, few answers were given to the problems of agricultural professionals. For the farmers, the PEPAC did not consider the specificities of the northern Portuguese region, small farmers, family farming, mountain farming and sectors experiencing cash flow difficulties or in need of investment.

2. Methodology

The proposed theme has a strong empirical component, so the information used (television reports, farmer movement manifestos, magazine/newspaper articles and opinion articles) was analyzed, synthesized, interpreted and decoded to build a plan that makes it possible to satisfy the farmers demands and their movements and maintain the essential course set by the PEPAC.

The objectives of carrying out this work were to synthesize and reflect on the reasons, arguments and demands made by the civic movements of farmers in the north region, who expressed their dissatisfaction with the current PAC (2023–2027). It should be noted that the focus of analysis is on occurrences recorded only in the NUT II north of Portugal, as these cover specific aspects of regional agriculture and are important for CCDR Norte to incorporate into CAP review proposals. However, the interpretation of the facts is transversal to regions and Member States, given that the main arguments come from the rules imposed above all by Regulation (EU) 2021/2116 of the European Parliament and of the Council of 2 December 2021.

In the analyzed period, four farmers protests occurred in the northern region, one on February 2nd (Lamego) with about 300 farmers [4], two on February 6th (Valença and Vila Real) with about 200 farmers each [5,6], and one on February 7th (Macedo de Cavaleiros/ Carrazeda de Ansiães) with 300 farmers [4].

3. Results and Discussion

The cut in support for eco-schemes was the reason for the greatest discontent expressed by Portuguese farmers. However, this reason was often associated with increased production costs, which meant that many farms had their economic viability called into question (Table 1). Without the ability to increase the prices of agricultural products, and with margins already strangled by distribution and the reduction in consumer purchasing power, the cuts in direct support were, for many farmers, the overflowing of the cup.

Unfair competition and lack of market regulation: One of the aspects most mentioned by farmers and even publicly recognized by the Prime Minister of the XXIV Constitutional Government lies in the entry of agricultural products from outside the European Union, which do not comply with the same agricultural production rules and workers' conditions and rights. Part of the problem with these products entering the market is the lack of regulation and the ease and economic interest of agri-food distribution and retail companies in acquiring these agricultural products in other geographies. Still regarding market regulation, farmers demand a greater balance of margins between production and distribution, reflecting in a more equitable and fair way the distribution of value, among all operators in the agri-food chains, and favoring dignified remuneration for farmers, for the goods they produce and the ecosystem services they provide.

Table 1. Aggregation of the reasons presented and summary of the origin of the problems.

Reasons	Source of the Problem
Cuts in subsidies	Reduction in payments for agri-environmental measures
Decrease in profitability	Increase in production costs (fuel, energy and animal feed)
Excessive bureaucracy	Increased requirements in animal registration and health
Excessive bureaucracy	Need to fill out field notebooks
Excessive bureaucracy	Difficulties in obtaining activity licensing
Excessive environmental demands	Increase in environmental rules with Farm to Fork and Green Deal
Subsidies and eligibility	Reduction in the eligibility of common land areas for CAP support
Subsidies and eligibility	Allocation of support based on surface area, which is damaging for small farms
Subsidies and eligibility	Need for support to implement anti-hail towers (Carrazeda de Ansiães)
Lack of market regulation	Value chain with unfair valuation of primary production
Lack of renewal of human capital	Low attractiveness of human resources for the sector
Unfair competition	Entry of competing products from outside the EU that do not respect the same rules

Source: [1,2,4–8].

Eligibility and cuts subsidies: Given the need for a more regionalized PEPAC, farmers, especially those who practice agriculture in mountain areas and small family farms, consider the distribution of support unfair, stating that their farms do not fit into the measures created and that they receive little or no benefit from the CAP. They emphasize that their agricultural production systems fall within environmentally friendly production systems, but due to the rules currently in force, they are not eligible for support. Regarding reasons associated with support, the main factor that triggered farmers' discontent was due to the cut in eco-schemes, namely organic farming (35%) and integrated production (25%), and delays in payment deadlines. The INE (2023) highlights a sharp decrease in other production subsidies (–47.3%) related to the payment schedule, which would last until June 2024.

Decrease in profitability: Farmers strongly emphasize that they continue to sell their products at low prices and that consumers purchase them in markets and large supermarkets at significantly higher prices. At the same time, production costs have increased (especially energy, fertilizers and concentrated food for cattle), unbalancing the treasury of their farms and putting at risk the livelihood and dignified life of them and their families.

Excessive environmental demands: The strategy associated with the European Ecological Pact (Green Deal), which aims at an ecological transition with the objectives of reducing CO_2 emissions by 55% by 2030 (compared to 1990 levels) and decarbonization by 2050, is contested by farmers who feel unfairly compared to other more polluting sectors, namely industry and transport, and have greater support for the ecological transition. As Rodríguez-Pose and Bartalucci (2023) [9] suggest, some regions are more exposed than others to the major shifts prompted by the Green Deal.

Lack of renewal of human capital: The renewal of human capital is associated with two aspects, on the one hand the attractiveness of the profession for young people and the necessary support for the installation of young farmers, on the other the valorization of the sector in society and the role of agriculture in supplying goods for consumption essential to human life and health.

Actions and Changes Introduced to Satisfy Demands

Between March 7th and April 8th, the European Commission invited farmers to respond to a survey on the administrative simplification of the CAP, in order to identify the sources of administrative burden and complexity arising from the rules, as well as other rules applicable to food and agriculture. Results were expected to be published by October 2024, clarifying the sources of complexity for farmers.

As this is a problem that affects a fundamental issue for society, human nutrition, the Government is committed to the stability of supplies and prices of agricultural products, simultaneously with the adaptation of national plans to initiatives and European strategies on agriculture, the environment, climate change and living and working conditions. It is natural that there will be new and rapid developments on the topic, and in this sense, new and important actions and solutions may emerge after the conclusion of this work.

After the meetings of the farmer civic movements and until the end of the mandate of the XXIII Constitutional Government, some actions and changes were implemented to respond to their demands (Table 2).

Table 2. Actions and changes introduced.

Reason	Action/Change
Excessive environmental demands	Partial derogation from the rules on fallow land (BCAA 8 standard)
Decrease in profitability	Credit line with subsidized interest of EUR 100 million (Treasury Line–agricultural sector) to support treasury costs to finance activity
Cuts in subsidies	Changes to payments for agri-environmental measures and eco-schemes, reinforcement of EUR 60 million
Inaccessibility to subsidies	Creation of a "*Temporary Compensation Measure*" to boost farmers' access to eco-schemes
Lack of market regulation	Meeting of the Platform for Monitoring Relations in the Agro-Food Chain (PARCA) 22 March 2024

4. Conclusions

Farmers, like what other professionals demand on the streets (teachers, doctors, nurses, security forces, etc.), want in the present context and time an improvement in their living and working conditions, with justice and a decent income. It is curious and welcomed that the spontaneous civic farmers movements, created from informal groups and organized through social networks in favor of a common good, organized themselves and raised public awareness and that of the government, placing the topic of agriculture and the rural world on the media agenda.

At a time when Portugal was in a pre-electoral campaign phase, without resorting to the organizations that represent them, these civic movements of farmers had a prominent and negotiating position, presenting directly to the Minister, who heard and sought to respond to their aspirations.

The environmental, social and economic importance of agriculture in Portugal, with the occupation of around 40% of the territory, a gross added value of 1.7% of Gross Domestic Product (GDP) and an agricultural population that represents 5.7% of the resident population [10], goes far beyond these numbers; it constitutes security for independence and sovereignty, a pillar of its cultural identity and indispensable interconnection with other sectors.

Author Contributions: B.L. and J.V. contributed to the conceptualization, implementation, analyses, writing and reviewing of this study. All authors have read and agreed to the published version of the manuscript.

Funding: This research received no external funding.

Data Availability Statement: The data presented in this study are public and therefore available in the open sources cited in the references.

Conflicts of Interest: The authors declare no conflict of interest.

References

1. Agricultores Portugueses Manifestam-Se Por "Condições Mais Justas". *Euronews*. Available online: https://pt.euronews.com/video/2024/02/01/agricultores-portugueses-manifestam-se-esta-quinta-feira-por-condicoes-mais-justas (accessed on 2 February 2024).
2. Porque É Que Os Agricultores Europeus Estão a Protestar? E Em Portugal? *Jornal Público*. Available online: https://www.publico.pt/2024/02/01/azul/noticia/agricultores-europeus-estao-protestar-portugal-2078879 (accessed on 1 February 2024).
3. Rodríguez-Pose, A.; Dijkstra, L.; Poelman, H. The geography of EU discontent and the regional development trap. *Econ. Geogr.* **2023**, *100*, 213–245. [CrossRef]
4. Agricultores do Distrito de Bragança Agendam Manifestação Para Quinta-Feira. *Observador*. Available online: https://observador.pt/2024/02/07/agricultores-do-distrito-de-braganca-agendam-manifestacao-para-quinta-feira/ (accessed on 7 February 2024).
5. Diana, M. Agricultores do Norte Vão Manter Bloqueios Em Valença Até Serem "Ouvidos" e Terem "Alternativas". *Jornal de Negócios*. 2024. Available online: https://www.jornaldenegocios.pt/empresas/agricultura-e-pescas/detalhe/agricultores-do-norte-vao-manter-bloqueios-em-valenca-ate-serem-ouvidos-e-terem-alternativas (accessed on 6 February 2024).
6. Protestos Dos Agricultores: Marcha Lenta Com Cerca de 100 Viaturas Bloqueia Acesso a Valença. *Jornal Expresso*. Available online: https://expresso.pt/economia/economia_agricultura/2024-02-06-Protestos-dos-agricultores-marcha-lenta-com-cerca-de-100-viaturas-bloqueia-acesso-a-Valenca-e72e13af (accessed on 6 February 2024).
7. Flor, A. No Movimento Civil de Agricultores Cabem Todos Os Descontentes. *Jornal Público*. 2024. Available online: https://www.publico.pt/2024/02/01/azul/noticia/movimento-civico-agricultores-cabem-descontentes-2079002 (accessed on 1 February 2024).
8. Segundo Dia de Protestos Dos Agricultores Portugueses. *RTP Notícias*. Available online: https://www.rtp.pt/noticias/pais/segundo-dia-de-protestos-dos-agricultores-portugueses_e1547829 (accessed on 2 February 2024).
9. Rodríguez-Pose, A.; Bartalucci, F. The green transition and its potential territorial discontents. *Camb. J. Reg. Econ. Soc.* **2023**, *17*, 339–358. [CrossRef]
10. INE. Contas Económicas da Agricultura 2023-1. ª Estimativa. Instituto Nacional de Estatística. 2023. Available online: https://www.ine.pt/xportal/xmain?xpid=INE&xpgid=ine_destaques&DESTAQUESdest_boui=593652342&DESTAQUESmodo=2 (accessed on 5 February 2024).

proceedings

MDPI

Proceeding Paper

Education on Sustainable Development Through Built Heritage and Technology: The "Art Nouveau Path" in the EduCITY App †

João Ferreira-Santos * and Lúcia Pombo

Research Centre on Didactics and Technology in the Education of Trainers (CIDTFF), Department of Education and Psychology (DEP), University of Aveiro (UA), 3810-193 Aveiro, Portugal; lpombo@ua.pt

* Correspondence: joaomrsantos@ua.pt

† Presented at the XVIII International Seminars on Overarching Issues of the European Area, Porto, Portugal, 23–26 May 2024.

Abstract: The value of cultural heritage for sustainable development has been acknowledged since the mid-20th century. Built heritage plays a key role in preserving identity, memory, and, currently, aiding education through mobile technology. The integration of education, heritage, and technology presents opportunities for innovation and originality, which are crucial for engaging students in sustainability. This paper focuses on the development of the "Art Nouveau Path" mobile game within the "EduCITY" app, which has been designed to promote Education for Sustainability. The game explores Aveiro's Art Nouveau heritage and its alignment with nature and environmental concerns.

Keywords: sustainable development; key competences for sustainability; EduCITY; mobile game; Art Nouveau built heritage; Art Nouveau Path

1. Introduction

In the context of the current global focus on sustainability, new entities are emerging, emphasising the significance of cultural heritage in fostering sustainable development.

The recognition of cultural heritage's importance, particularly in preserving identity and memory, has grown since the mid-20th century [1]. Technological advancements, particularly in mobile technologies, have expanded the potential for exploiting cultural heritage for educational purposes. This is exemplified by the development of "Art Nouveau Path", a Mobile Augmented Reality Game (MARG) in the "EduCITY" application, which aims to enhance Education for Sustainability through Aveiro's Art Nouveau built heritage.

This paper is divided into four sections. The first and second, Introduction and Methodology, respectively, inform the reader about the paper's structure and research methods. The third section is devoted to contextualising this research and the development of the MARG "Art Nouveau Path". The final sections, which constitute a kind of conclusion, set out the future work that will be carried out.

2. Methodology

Considering the nature of this paper, a narrative review of the literature was conducted. This comprehensive review had a main purpose: to create a knowledge base capable of inspiring and guiding future work in these specific thematic areas.

3. Context

3.1. The EduCITY Project

The EduCITY Research and Development project (https://educity.web.ua.pt/, accessed on 6 May 2024) is grounded on the premise that sustainable cities and communities can be developed through the use of digital technologies as agents of change, with the aim of fostering awareness of issues concerning sustainable development. In this context, the EduCITY project aims to promote the urban space as a space for sustainable dynamics by facilitating a smart educational city environment using a mobile application (https://educity.web.ua.pt/app.php, accessed on 6 May 2024), which stores mobile games based on challenges that include, among other educational multimedia content, educational content in augmented reality (AR).

The EduCITY games can be developed and played collaboratively by the school public, as well as by locals and even tourists. These mobile educational games comprise an exploration of the surrounding environment through an interdisciplinary lens, which is facilitated by pre-defined routes that allow users to explore the urban space and beyond. One of EduCITY's potential features is the ability for any citizen to co-create games and educational resources, including those in AR. In this way, in addition to fostering sustainability literacy, EduCITY also advocates for the advancement of digital literacy and skills. This is predicated on the engagement of individuals with the objective of actualizing the need to change habits and awareness towards effective sustainable development [2,3].

3.2. The Role of Built Heritage in Education

The concept of built heritage encompasses tangible heritage, such as historic buildings or structures and cultural sites, including heritage landscapes. It provides a source and testimony about human history and development [4]. However, it should be noted that built heritage not only represents the past, but also the present, and is configured as a bridge to the future. Consequently, built heritage enables the exploration of various dimensions of human development, encompassing cultural and artistic perspectives, technical considerations, and ethical and social dimensions [4]. These diverse interpretations offer a potential resource for educational purposes, including direct applications in History and Art History, Citizenship, and Education for Sustainable Development.

4. The Mobile Game "Art Nouveau Path"

The development of an MARG involves the integration of diverse areas of knowledge, with an emphasis on the significance of contextual learning. This approach is crucial for a comprehensive understanding of complex phenomena, including those related to sustainability and sustainable development. In this context, it is vital to adopt a multifaceted perspective when examining the myriad phenomena that surround us, leveraging existing resources, such as built heritage. This will demonstrate the potential for collaboration with other fields and areas of knowledge.

4.1. The Role of Art Nouveau in Education for Sustainability

Art Nouveau, as an artistic movement, thrived during a relatively brief period between the late 19th and early 20th centuries. This style is often regarded as a comprehensive art form that encompassed various media, including architecture, painting, sculpture, decorative arts, graphic arts, textiles, fashion, and jewellery [5,6].

Art Nouveau emphasised a deep connection to nature as a counterbalance to the increasingly mechanised world. According to Walter Benjamin, Art Nouveau interiors provided a sanctuary from the rapid urbanisation and noise of cities while also seeking to blur the distinction between art and technique. This movement utilised technical aspects to elevate artistic forms, resulting in stylised designs that could serve both functional and decorative purposes [7].

The defining characteristics of Art Nouveau products include flowing lines reminiscent of natural elements such as water and plants, as well as motifs inspired by fauna and representations of femininity. Moreover, the regional variations in Art Nouveau reflect the influence of local environments on artistic creations. This thematic and formal diversity underscores the movement's relevance to contemporary sustainability issues, emphasising the significance of local environments as sources of inspiration and the fusion of academic expertise with traditional craftsmanship [5,6].

The collaboration of local artisans and artists enabled the integration of academic knowledge from disciplines such as architecture, engineering, and sculpture with indigenous skills and techniques. This blending of expertise contributed to the distinctive semantic and visual grammar of Art Nouveau products, setting them apart from those of other contemporary styles such as Art Deco.

These connections allow us to state that Art Nouveau, as an artistic movement, permitted the inclusion of various strands, or 'regionalisms', which, while enriching the stylistic variety, were founded upon the relationship between the 'object produced', the local environment, and the various agents of the same environment where it was produced. This not only addresses the issue of environmental sustainability but also the holistic concept of sustainability, as recognised by all, in relation to the interconnectivity between the environment, society, and the economy.

4.2. The "Art Nouveau Path" Towards Education for Sustainability

The MARG "Art Nouveau Path" employs AR content in conjunction with other multimedia content and has yet to be implemented. This game is predicated on the premise that built heritage can serve as a space for learning and training [1]. Furthermore, it is based on the recognition of the potential of AR in the context of mobile learning and situated learning [8]. In addition to these considerations, it is important to note that built heritage plays a vital role in developing key competences for sustainability-related education.

The MARG "Art Nouveau Path" incorporates challenges that integrate AR educational resources, simulations, 3D animations, and various information points. The façades of buildings and other monuments become natural markers, allowing users to walk through the streets and squares of Aveiro's historic city centre and discover the educational content that these buildings 'hide' using a mobile device with the EduCITY app installed.

The development of this MARG adheres to the principles of design-based research and encompasses multiple iterations of improvement. The game is currently undergoing evaluation by a panel of experts in the fields of education and multimedia. The research and creation of this game are part of a doctoral research project whose objective is to contribute to the advancement of key competencies for Education for Sustainability through the valorisation of local built heritage [8].

To achieve this objective, the game will be implemented through a sequential explanatory quasi-experimental research plan, which will involve eighth-grade classes. These classes will be divided into two groups: a pre-test group and a post-test group. Questionnaires will be provided to both groups, and perceptions will be gathered through focus-group interviews.

4.3. The Potential of the "Art Nouveau Path" for Curriculum Development and Key Competences for Sustainability

Integrating the "Art Nouveau Path" into the educational curriculum requires aligning it with learning aims across various school subjects. The game allows for the exploration of historical, artistic, mathematical, and science content, among others. This is ensured by the game dynamics, which are grounded in quizzes. All questions have four answer choices and there is feedback on the chosen answer, even if it is correct. This makes it possible to add more information through the feedback. Furthermore, the "Art Nouveau Path" encourages interdisciplinary learning, allowing students to make connections between different subjects and see the broader implications of their studies. This holistic approach to education can foster a more integrated understanding of sustainability, emphasising the interconnectedness of cultural, historical, and environmental factors on the development of sustainability key competences [4].

The educational resources provided within the "Art Nouveau Path" are also designed to foster critical thinking and problem-solving skills. These resources comprise interactive quizzes, historical narratives, and real-time feedback, all delivered through the interface. By engaging with these elements, students can gain a deeper understanding of the principles of sustainable development and the role of cultural heritage in supporting these principles [8].

5. Conclusions

Since the "Art Nouveau Path" is still under development, it is hoped that built heritage will be recognised and valued as a new agent for Education for Sustainable Development. This will allow for the implementation of new curricular approaches, which will benefit from the local context of the heritage itself in conjunction with the potential of learning through mobile augmented reality. In this way, the significance of individual and collective memory and identity will be emphasised, thereby fostering awareness of the necessity for sustainable development to construct a healthier, fairer, and safer world for all.

Author Contributions: Conceptualization, J.F.-S. and L.P.; investigation, J.F.-S.; resources, J.F.-S.; writing—original draft preparation, J.F.-S.; writing—review and editing, J.F.-S. and L.P.; supervision, L.P. All authors have read and agreed to the published version of the manuscript.

Funding: The EduCITY project is funded by National Funds through the FCT—Foundation for Science and Technology, I.P., as part of the PTDC/CED-EDG/0197/2021 project. The first author's work is funded by National Funds through FCT—Foundation for Science and Technology, I.P., within the scope of a Doctoral Research Grant with the reference 2023.00257.BD.

Institutional Review Board Statement: Not applicable.

Informed Consent Statement: Not applicable.

Data Availability Statement: Data sharing is not applicable to this article.

Conflicts of Interest: The authors declare no conflicts of interest.

References

1. Council of Europe. *European Charter of the Architectural Heritage*; Council of Europe: Strasbourg, France, 1975.
2. Pombo, L. Exploring the role of mobile game-based apps towards a smart learning city environment—The innovation of EduCITY. *Educ. Train.* **2022**, *65*, 253–264. [CrossRef]
3. Redman, A.; Wiek, A. Competencies for Advancing Transformations Towards Sustainability. *Front. Educ.* **2021**, *6*, 785163. [CrossRef]
4. Smith, L. *Uses of Heritage*; Routledge: London, UK, 2006.
5. da Cidade de Aveiro, M. (Ed.) *Dicionário Arte Nova*; Câmara Municipal de Aveiro: Aveiro, Portugal, 2011.
6. Sembach, K.-J. *Arte Nova*; Taschen: Cologne, Germany, 2010.

7. Benjamim, W.S. Pintura, Arte Nova [Judendstil], Novidade. In *As Passagens de Paris*; Assírio & Alvim: Lisboa, Portugal, 2019; pp. 674–694.
8. Van Doorsselaere, J. Connecting sustainable development and heritage education? An analysis of the curriculum reform in Flemish public secondary schools. *Sustainability* **2021**, *13*, 1857. [CrossRef]

proceedings

MDPI

Proceeding Paper

The Role of Parish Councils in Georeferencing for Emergency Planning: Approach and Methods †

Andreia Rodrigues [1,*], Aldina Santiago [1], Domingos Xavier Viegas [2] and José Luís Zêzere [3]

1 ISISE-UC Institute for Sustainability and Innovation in Structural Engineering, Department of Civil Engineering, University of Coimbra, Rua Luís Reis Santos, Pólo II, 3030-788 Coimbra, Portugal; aldina@dec.uc.pt
2 ADAI-Association for the Development of Industrial Aerodynamics, Department of Mechanical Engineering, University of Coimbra, Rua Luís Reis Santos, Pólo II, 3030-788 Coimbra, Portugal; xavier.viegas@dem.uc.pt
3 Institute of Geography and Spatial Planning, University of Lisbon, 1649-004 Lisboa, Portugal; zezere@campus.ul.pt
* Correspondence: andreia.rodrigues@dec.uc.pt; Tel.: +351-916009416
† Presented at the XVIII International Seminars on Overarching Issues of the European Area, Porto, Portugal, 23–26 May 2024.

Abstract: Rural fires are becoming increasingly impactful, not only due to their rapid and intense spread but also because of the growing number of people affected. In Portugal, numerous remote housing clusters are difficult to access and are primarily occupied by seniors or individuals with disabilities. These areas are situated in high-risk wildfire zones. Therefore, it is crucial to document the locations of the individuals who are most at risk to ensure that relief efforts can be focused on and directed towards them in case of emergency. The aim of this paper is to propose an inventory plan for communities at risk of wildfire, requiring the involvement of parish councils. The plan involves the use of simple and cost-free georeferencing tools (such as Mymaps of Google) to enable these organizations to assist in municipal emergency planning and facilitate the exchange of real-time information. The outcomes of this paper will be demonstrated in three parishes in mainland Portugal, where the key benefits and limitations were identified. The goal is to enhance the project for potential replication in other areas.

Keywords: wildfires; parish councils; inventory; vulnerable population

1. Introduction

Several climate studies now reveal a worrying trend, especially in Portugal and other southern European countries, regarding the occurrence of extreme natural events. Research carried out as part of a broader study by the European Forestry Institute on forest fires predicts that temperature levels (especially in summer) in Portugal will be higher and precipitation levels lower than average [1]. The rise in temperatures and the decrease in humidity are well-established trends, showing no signs of abating soon and appearing to be key drivers of major fires in the coming years [2].

In Portugal, many isolated housing clusters are difficult to access and predominantly inhabited by the elderly or people with disabilities and are in regions with a high risk of wildfire. The problem of rural abandonment is widely recognized, especially in the northern and central parts of the country, with villages becoming increasingly isolated and populations aging, and where the forest increasingly enters the villages, as the residents no longer have the capacity to practice agriculture or clear fuel around the villages due to their advanced age [3].

The aforementioned factors, combined with the sociodemographic changes experienced in Portugal, result in an increasing number of inhabitants being exposed to the danger of rural fires [4]. An analysis of the victims of wildfires in Portugal in 2017, focusing on their behavior when the fires approached, revealed that 25% of the victims died inside their homes, having been caught by the fires, without receiving an alert or evacuation order [5]. In this way, it becomes imperative to know the locations and those with limitations in perception and reaction, so that authorities can, in advance, evacuate these people from dangerous zones, as they may not be able to do so on their own.

2. The Role of Parish Councils in Georeferencing

Emergency management primarily involves handling the period between the emergency and the arrival of structured assistance, to safeguard the safety of people and property in the affected area, thereby limiting the negative consequences of the event [6].

Parishes are local authorities in Portugal that, within the municipal territory, aim to serve the interests of the population residing in each parish area [7]. There are currently 3091 parishes: 2882 parishes in mainland Portugal, 155 in the Autonomous Region of the Azores and 54 in the Autonomous Region of Madeira. Parish councils are crucial in the processes of governance and public management. These governance structures are the entities closest to the population, allowing them to more easily identify the problems faced by the most vulnerable individuals. They can also contribute to maintaining an updated database, which can be used in emergencies to support decision-making regarding necessary actions.

3. Approach

Considering the awareness raised above, a methodology was developed involving parish councils, where individuals working within this structure could actively contribute to maintaining an updated database of the most vulnerable people and those with special needs who require special attention during natural disasters.

3.1. Steps

The identification and analysis of population vulnerabilities are fundamental in the current scenario of climate change, both for preventing and preparing for disasters and for facilitating relief during emergencies. To facilitate the inventory process of the most vulnerable population and those with special needs, three essential steps were established:

- Identification of clusters at risk: In this step, local authorities identify the highest priority villages to focus on, based on the type of population, the village's surroundings, and its accessibility.
- Survey of the vulnerable population: Visits are carried out to the villages, with purposes such as monitoring the local inhabitants, finding where the inhabited houses are and surveying to identify the type of inhabitants and recording of any special need that must be taken into account in case of an emergency.
- Processing and making data available: All collected information is entered into Google's My Maps, with only data relevant to emergency planning being recorded.

The simple step-by-step application scheme allows us to identify some implementation advantages that allow for greater acceptance by those involved (Figure 1).

After identifying the villages at greatest risk, an on-the-ground survey of people with physical and mental limitations is carried out, and these are recorded on a grid but also marked in the Google My Maps tool. This tool provides easy and free access, requiring only internet access and a smartphone for the operator. The application only records the location of the person with limitations, their housing conditions, and any special needs, without adding any personal information or sensitive data. Once the database is established, authorized users can add or update information at any time. All this information can be printed off or consulted offline on a mobile phone without the need to install any software or use complex GIS tools.

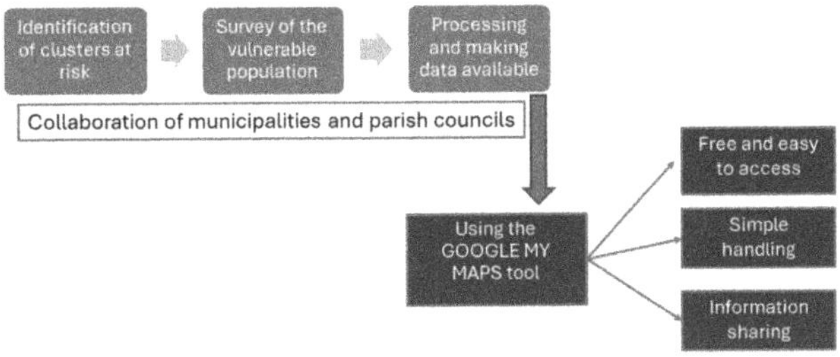

Figure 1. Inventory procedure.

3.2. Test Group

After organizing the steps, this procedure was tested in four parishes in the center of Portugal, in the municipalities of Lousã, Miranda do Corvo and Anadia. Meetings were held with parish council representatives to discuss the added value and challenges of the procedure. The aim is to use this system for a year to evaluate its advantages.

Following the surveys, in which the local population was involved, council staff updated the database whenever there were changes in occupation, e.g., death of a resident or new inhabitants. As the population felt involved and was satisfied with these procedures, the village inhabitants themselves began alerting the council to changes in occupation, becoming important players in the entire process. The information remains constantly updated in platform (Figure 2), and in two cases, it has been tested through a simulation exercise.

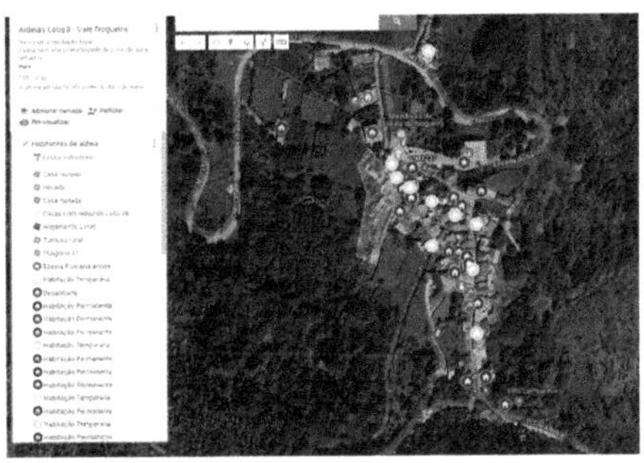

Figure 2. Graphic representation in MyMaps by Google of the village of Vale Nogueira, with all georeferenced inhabitants.

In this exercise, it was observed that information was easily shared with operational decision-makers, facilitating decision-making and resource planning. When a large fire approaches, the command post provides real-time information, which can be integrated into decision support systems, thus being used in decisions such as the need to evacuate a village.

After a year of use, those responsible for the parish councils were interviewed and were invited to present the advantages and limitations of the procedure. The parish councils identified the following advantages:

- The ease of use of the application motivated them to use it in other villages.

- Georeferencing the most vulnerable people and those with special needs becomes important, not only for forest fire situations but also for other relief needs.
- The ease of sharing this information allowed, for example, the improvement of patrolling actions by security forces.

The main constraints were identified:

- There are mobile network difficulties in some areas of the interior of Portugal—in some situations, data insertion and updating could not be carried out in the village itself. This same difficulty could compromise the sending of information to the command post in a real fire situation.
- In larger parishes, to keep the database up to date, it was suggested that each village have a liaison officer.

4. Conclusions

Portugal faces a problem with its population, especially in the interior regions. Many villages are scattered in areas of high fire risk, and mostly inhabited by elderly people or those with physical and cognitive limitations. Therefore, it is important that the rescue services know exactly where these people are and the conditions in which they can be found.

One of the biggest challenges is keeping this information updated over time. By using and valuing the proximity of parish councils, it is possible to maintain an updated database, using simple and free tools. These tools allow information sharing and support civil protection services, which often lack the means and resources to keep this information updated.

Parish councils, as the government structure closest to the people, play a fundamental role in maintaining a population database, using simple georeferenced systems. The use of complex georeferencing tools has distanced local authorities from the inventory process, and this process presents an attempt to solve this problem, creating a simple system that only requires a smartphone and little specific training.

With the dissemination of this procedure, it is hoped to gradually achieve an inventory of the territories most exposed to extreme dangers, so that their safety is ensured before and during a possible damaging event.

Author Contributions: Conceptualization, A.R. and D.X.V.; methodology, A.R.; software, A.R.; validation, A.S., D.X.V. and J.L.Z.; writing—original draft preparation, A.R.; writing—review and editing, A.R.; visualization, D.X.V. and J.L.Z.; supervision, A.S.; project administration, A.S.; funding acquisition, A.S. All authors have read and agreed to the published version of the manuscript.

Funding: This work is financed by national funds through FCT—Foundation for Science and Technology under grant agreement 2023. 02015.BD attributed to the 1st author and project "EVACUARFLORESTA—decisions and evacuation plans in forest fire scenarios", grant number PCIF/AGT/0061/2019.

Institutional Review Board Statement: Not applicable.

Informed Consent Statement: Not applicable.

Data Availability Statement: Data sharing is not applicable to this abstract.

Conflicts of Interest: The authors declare no conflicts of interest.

References

1. Moreira, F.; Catry, F.X.; Rego, F.; Bacao, F. Size-dependent pattern of wildfire ignitions in Portugal: When do ignitions turn into big fires? *Landsc. Ecol.* **2010**, *25*, 1405–1417. [CrossRef]
2. Beighley, M.; Hyde, A.C. *Gestão dos Incêndios Florestais em Portugal numa Nova Era Avaliação dos Riscos de Incêndio, Recursos e Reformas*; Center for Forestry Studies—Higher Institute of Agronomy of the University of Lisbon: Lisbon, Portugal, 2018.

3. Manzello, S.L.; Blanchi, R.; Gollner, M.J.; Gorham, D.; McAllister, S.; Pastor, E.; Planas, E.; Reszka, P.; Suzuki, S. Summary of workshop large outdoor fires and the built environment. *Fire Saf. J.* **2018**, *100*, 76–92. [CrossRef] [PubMed]
4. Molina-Terrén, D.M.; Xanthopoulos, G.; Diakakis, M.; Ribeiro, L.; Caballero, D.; Delogu, G.M.; Viegas, D.X.; Silva, C.A.; Cardil, A. Analysis of forest fire fatalities in Southern Europe: Spain, Portugal, Greece and Sardinia (Italy). *Int. J. Wildland Fire* **2019**, *28*, 85–98. [CrossRef]
5. Rodrigues, A.; Santiago, A.; Laím, L.; Viegas, D.X.; Zêzere, J.L. Rural Fires—Causes of Human Losses in the 2017 Fires in Portugal. *Appl. Sci.* **2022**, *12*, 12561. [CrossRef]
6. Alexander, D.; David, E. *Principles of Emergency Planning and Management*; Oxford University Press: Oxford, UK, 2002.
7. Freitas do Amaral, D. *Administrative Law Course—Volume I*; Livraria Almedina: Lisbon, Portugal, 1986.

proceedings

MDPI

Proceeding Paper

Enhancing Climate Resilience Through Aligned Municipal Strategies †

Madalena Cabeleira [1] and Ana Paula Oliveira [1,2,*]

1 Instituto Superior de Educação e Ciências (ISEC Lisboa), Alameda das Linhas de Torres, 179, 1750-142 Lisboa, Portugal; 20210445@alunos.iseclisboa.pt
2 Centro de Investigação, Desenvolvimento e Inovação em Turismo (CiTUR)—Polo do Estoril, Avenida Condes de Barcelona, n.° 808, 2769-510 Estoril, Portugal
* Correspondence: ana.oliveira@iseclisboa.pt
† Presented at the XVIII International Seminars on Overarching Issues of the European Area, Porto, Portugal, 23–26 May 2024.

Abstract: Cities are increasingly vulnerable to climate change, prompting many municipalities to adopt Municipal Strategic Plans for the Adaptation to Climate Change (MSPACC). The success of these plans depends on coordinated efforts, particularly from Municipal Civil Protection Services (MCPS), which play a key role in climate adaptation. To assess their involvement, a survey of MCPS coordinators was conducted, yielding 279 responses (90.6% of all Portuguese MCPS). Statistical analysis revealed that 53.4% were familiar with adaptation strategies, with 59.1% identifying MSPACC development and implementation as a key approach. Additionally, 63.4% acknowledged that climate risks were anticipated in their Municipal Civil Protection Emergency Plans (MCPEP). However, findings highlighted a critical misalignment between MSPACC and MCPEP, potentially weakening both adaptation and mitigation efforts. Bridging this gap through integrated planning and collaboration is essential to strengthening municipal resilience and safeguarding communities.

Keywords: climate changes; vulnerability; resilience; municipal climate change adaptation plans; civil protection

1. Introduction

Rapid urbanization and climate change are intensifying the challenges faced by modern cities, particularly in relation to climate-related disasters [1]. As a result, urban areas are becoming increasingly vulnerable, requiring urgent action and strategic planning to mitigate risks and enhance resilience.

In response, many municipalities have adopted Municipal Climate Change Adaptation Plans (MCCAP) or Municipal Climate Change Adaptation Strategies (MCCAS), collectively herein referred as Municipal Strategic Plans for the Adaptation to Climate Change (MSPACC). These frameworks aim to enhance resilience by guiding local governments in mitigating risks and adapting to evolving climate threats [2]. However, municipalities contribute to climate adaptation in more extensive and multifaceted ways beyond MSPACC, including the integration of social development models and broader public policy approaches. While this article focuses on MSPACC and their relationship with Municipal Civil Protection Services (MCPS), further research should explore additional municipal contributions to climate adaptation, such as infrastructure development, urban planning, and public health initiatives.

The success of MSPACC relies on strong collaboration among various stakeholders, including the general population, public and private sectors, and MCPS. These services

https://doi.org/10.3390/proceedings2025113019

play a crucial role in ensuring public safety and enhancing resilience against climate threats. The "Resilient Cities in Portugal" campaign further supports these efforts by emphasizing risk prevention, proactive threat anticipation, effective disaster response, and rapid recovery [3,4]. Cities that embrace smart and adaptive strategies tend to be more resilient, capable of responding swiftly to crises while promoting the well-being of their inhabitants [5,6].

For MSPACC to be effective, well-structured policies must be complemented by active engagement from all sectors, ensuring that adaptation measures are practical, sustainable, and widely understood.

This study aimed to assess whether the adaptation plans and strategies implemented by municipalities serve as effective tools for preventing accidental events and disasters linked to climate change (CC). By evaluating their role in risk mitigation and emergency preparedness, this research contributes to understanding how municipalities can better integrate climate adaptation into local governance, ultimately strengthening community resilience and disaster prevention efforts.

2. Methodology

A survey was conducted using Google Forms®, consisting of 16 questions—13 closed-ended and 3 open-ended—divided into two main sections: (i) Sociodemographic data and (ii) information on climate change at the municipal level.

The survey was distributed between 24 October 2022, and 9 June 2023, via email and social media, to coordinators of the Portuguese MCPS. In some cases, follow-up phone calls were made to encourage participation.

A total of 279 responses were collected, representing 90.6% of all MCPS in Portugal. The data were analyzed using Microsoft Excel® version 28 for Windows and the Statistical Package for the Social Sciences® (SPSS) version 25 for Windows. Statistical analysis included descriptive statistics (absolute and relative frequencies) and inferential statistics, specifically the Chi-square test of independence.

Additionally, a geographical and territorial breakdown of the surveyed municipalities was conducted to identify regional trends in adaptation strategies.

3. Results and Discussion

The analysis of the collected data highlights key insights into the awareness and implementation of CC adaptation strategies among MCPS coordinators. When asked about the existence of any adaptation strategy in their municipality, 53.4% of coordinators reported being aware of such measures. Among those who acknowledged the presence of an adaptation strategy, 59.1% specifically referenced the development and implementation of an MSPACC, while 13.4% mentioned sustainability initiatives, and 6.0% cited awareness-raising actions (Figure 1).

A critical issue in the interpretation of these responses is the lack of a clear definition of climate adaptation actions and strategies. Some respondents identified awareness campaigns as adaptation strategies, despite these being supplementary actions rather than comprehensive policy instruments. This confusion underscores the need for clearer guidelines on what constitutes climate adaptation.

Regarding climate risks, coordinators were asked whether their Municipal Civil Protection Emergency Plans (MCPEP) accounted for the increasing threats posed by CC. The results showed that 63.4% confirmed their plans included such considerations, while 35.2% stated they did not, and 1.4% were unaware. Additionally, when asked whether their MCPEP incorporated strategic actions to address climate-related threats, responses were nearly identical. However, it remains uncertain whether these responses reflect formal policy adoption, actual implementation, or subjective perceptions lacking scientific validation. This limitation underscores the need for further empirical research to evaluate the true integration of climate risks into MCPEPs. Despite these uncertainties, municipalities can enhance the alignment between adaptation plans (MSPACC) and emergency protocols (MCPEP), ensuring local governments are better equipped to mitigate risks and protect their communities from the escalating impacts of climate change.

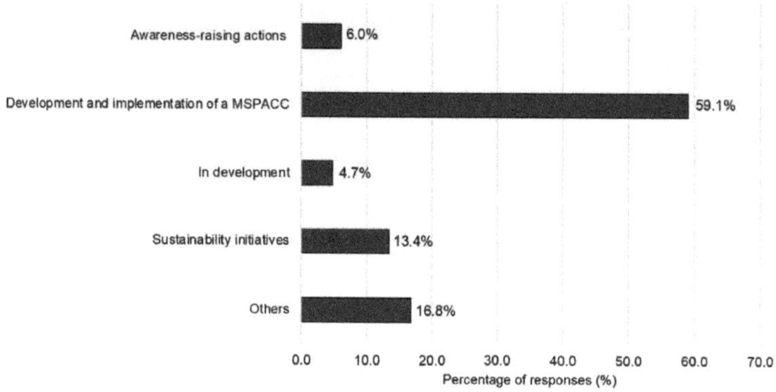

Figure 1. Survey responses on adaptation strategies developed by the municipality.

When asked about the necessity of developing strategic actions to address CC, an overwhelming 97.8% of municipal coordinators responded affirmatively, with only 2.2% in opposition. This strong consensus highlights the growing recognition of CC as a critical challenge but also underscores the need for deeper knowledge and a more structured approach to implementation. However, opinions were divided regarding the effectiveness of existing strategic actions in reducing disaster risks—50.2% of municipalities reported positive outcomes, while 49.8% remained skeptical. These findings may obscure broader trends, as the survey reflects the perceptions of municipal civil protection coordinators, whose responsibilities are primarily tactical and operational rather than strategic. Consequently, their insights may not fully represent the municipality's overarching climate adaptation strategy, which is shaped by political leadership. Future research should incorporate perspectives from municipal policymakers to provide a more comprehensive evaluation of climate adaptation efforts.

Regional analysis (Figure 2) further reveals significant variations in climate adaptation strategies. Among the surveyed MCPS, 70% of coastal areas had an MSPACC in place, compared to just 45% in inland regions. Likewise, urban municipalities were more likely to integrate climate risks into MCPEP (68%) than rural ones (52%). These disparities highlight the influence of geography on adaptation approaches, with coastal and urban areas demonstrating a stronger commitment to proactive planning.

Overall, the findings suggest that climate adaptation strategies remain in their early stages and require further refinement, evaluation, and consolidation to enhance their long-term effectiveness. Strengthening collaboration between operational and strategic decision-makers will be essential in advancing more cohesive and impactful adaptation policies.

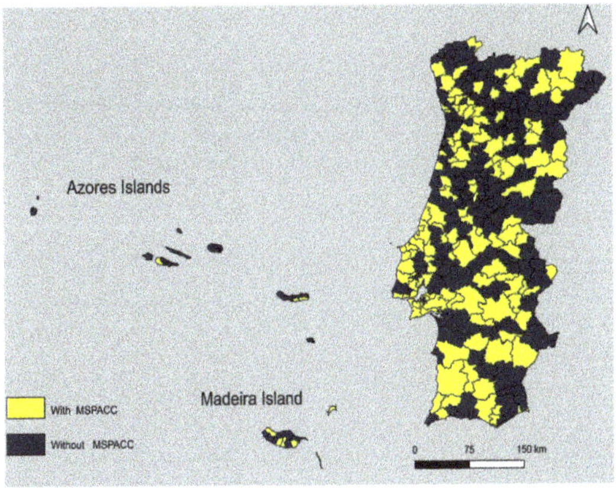

Figure 2. Geographical distribution of municipalities with and without MSPACC.

4. Conclusions

While interpreting the survey results as representative of municipalities' official positions carries the risk of bias, it is evident that addressing the growing impacts of CC requires concerted efforts at both municipal and community levels to ensure the well-being of local populations. The principle of "Think globally, act locally" serves as a powerful framework to empower communities by fostering awareness, resilience, and proactive adaptation strategies. Strengthening collaboration between municipalities and educational institutions can enhance learning, innovation, and knowledge-sharing, contributing to safer and more dignified living conditions. Integrating safety culture training within MSPACC can further build resilience by aligning preventive actions with the potential of educational cities.

However, research indicates that there is still a significant gap in establishing a strong safety culture in climate adaptation. Adaptation plans need to be more practical, flexible, and dynamic to ensure their effective implementation and responsiveness to evolving challenges. Self-protection measures, identified as a central theme, play a crucial role in embedding safety culture within climate narratives, empowering communities, and supporting civil protection efforts.

Building a resilient society requires active and sustained engagement from all sectors, promoting a widely integrated safety culture that strengthens collective adaptation and response capacities. While MCPS coordinators in Portugal recognize extreme weather events as key threats and view the MCPEP as an essential operational tool, its alignment with the MSPACC remains limited. Addressing these gaps through improved knowledge, coordination, and practical implementation of climate adaptation strategies is vital to enhancing municipal resilience and preparedness.

Author Contributions: Conceptualization, M.C. and A.P.O.; methodology, M.C.; formal analysis, M.C.; investigation, M.C. and A.P.O.; writing—original draft preparation, M.C.; writing—review and editing, M.C. and A.P.O.; supervision, A.P.O. All authors have read and agreed to the published version of the manuscript.

Funding: This research received no external funding.

Institutional Review Board Statement: The study was conducted in accordance with the Declaration of Helsinki and approved by the Ethics Committee of the Instituto Superior de Educação e Ciências (ISEC Lisboa) (Ethics Declaration of 15 September 2022).

Informed Consent Statement: Informed consent was obtained from all subjects involved in the study.

Data Availability Statement: The data presented in this study are available on request from the corresponding author due to the interviewee's privacy.

Acknowledgments: The authors would like to thank the coordinators of the Municipal Civil Protection Services of Portugal who responded to the survey. Thanks are also due to Raul Oliva for help in preparing Figure 2.

Conflicts of Interest: The authors declare no conflicts of interest.

References

1. Da Silva, M.F.; Moldero, L.S.; Trindade, L.H.; Barbosa, A.P. Smart cities and disaster prevention: Turning data into resilient strategies. *ARACÊ* **2024**, *6*, 18618–18631. [CrossRef]
2. European Union (EU). EU Adaptation Strategy. 2021. Available online: https://climate-adapt.eea.europa.eu/en/eu-adaptation-policy/strategy (accessed on 30 January 2025).
3. UN. *How To Make Cities More Resilient—A Handbook For Local Government Leaders*; United Nations International Strategy for Disaster Reduction (UNISDR); United Nations: Geneva, Switzerland, 2012; Available online: https://www.unisdr.org/files/26462_handbookfinalonlineversion.pdf (accessed on 15 December 2023).
4. UNDRR. History of Making Cities Resilient. United Nations Office for Disaster Risk Reduction (UNDRR). 2024. Available online: https://mcr2030.undrr.org/who-we-are/history (accessed on 4 February 2025).
5. Bibri, S.E. Data-driven smart sustainable cities of the future: An evidence synthesis approach to a comprehensive state-of-the-art literature review. *Sustain. Futures* **2021**, *3*, 100047. [CrossRef]
6. UNDP. Smarter and Inclusive Cities. Arup, Climate-KIC and UNDP. 2024. Available online: https://www.undp.org/sites/g/files/zskgke326/files/2024-06/smarterandinclusivecitiescourse_2.pdf (accessed on 4 February 2025).

proceedings

MDPI

Proceeding Paper

Railway Tourism and Sustainability in Águeda †

Eunice Duarte [1,*], Diogo M. Pinto [2] and Helena Pina [2]

[1] Centre for Research, Development and Innovation in Tourism (CiTUR-Leiria), School of Administrative Sciences, Polytechnic Institute of Lusophone Countries, 1700-098 Lisboa, Portugal
[2] Center for Studies in Geography and Spatial Planning (CEGOT), Faculty of Arts and Humanities, University of Porto, 4150-564 Porto, Portugal; dspinto@letras.up.pt (D.M.P.); mpina@letras.up.pt (H.P.)
* Correspondence: eunice.duarte29@gmail.com

† Presented at the XVIII International Seminars on Overarching Issues of the European Area, Porto, Portugal, 23–26 May 2024.

Abstract: The growing demand for sustainable tourism is evident in regions like Águeda, Center of Portugal, where the Historic Vouga Train attracts a significant number of visitors. This study analyzes the intersection of this historic train's use and sustainability principles, focusing on tourist demand. The methodology includes analyzing news from 2020 to 2023 and CP's passenger data. The findings show the train boosts tourism in Águeda, attracting visitors keen on historical and cultural preservation. Sustainable initiatives such as promoting public transport and minimizing environmental impact are highlighted. This study underscores the importance of integrating railway heritage preservation with sustainability goals to foster local economic development.

Keywords: Historic Vouga Train; railway tourism; sustainable tourism

1. Introduction

In the 19th century, the railway reduced travel time, enabling people to use trains for leisure. However, in the 1980s and 1990s, the railway system in Portugal underwent changes, with lines being closed or reduced. The Vouga line and the Aveiro branch were no exception, and in 1972, the section between Viseu and Sernada do Vouga was closed on the grounds that the steam locomotives caused forest fires [1,2]. Nevertheless, the population held a special affection for the Vouga Valley line, affectionately naming the train "Vouguinha" [3], a name still used by the populace today. Despite this, Comboios de Portugal (CP) decided not to use this name commercially and has demonstrated, as seen on the company's website, an effort to completely sever ties with this popular name, opting instead to use the commercial name Vouga Historic Train (Comboio Histórico do Vouga).

Águeda, a municipality in Center of Portugal, stands out for the historical and cultural value of its railway heritage through the workshops in Sernada do Vouga, the existence of a railway museum nucleus in Macinhata do Vouga, initiatives at Águeda station, and an old locomotive placed in the city. Moreover, Águeda has the potential to attract visitors through the Vouga Historic Train, a railway service offering a nostalgic journey aboard a centenarian locomotive. This is the only narrow-gauge line still in operation in Portugal [4], promoting the preservation of national, cultural, and historical railway heritage by repurposing it for tourism [5]. This tourism experience aligns with a growing interest in more sustainable tourism practices, reflecting a significant shift in traveler behavior towards combining leisure activities with environmental responsibility [6–8]. The railway, as a tourism resource, has been increasingly discussed, particularly for generating economic dynamism and differentiating localities through railway tourism [9,10]. In this context, the authors deemed

https://doi.org/10.3390/proceedings2025113020

it important to analyze how the use of the Vouga Historic Train relates to the principles of sustainability in the tourism demand of the Águeda region.

2. Materials and Methods

To achieve this objective, the methodology used was based on the analysis of news published by the media between 2020 and 2023, which covered the editions of the Vouga Historical Train. This analysis included journalistic coverage of the tourist impact and the frequency of news, in addition to data provided by CP regarding the number of passengers and the editions made to evaluate the impact of the train on tourist demand and on the sustainable development of the municipality of Águeda.

3. Results

Between 2020 and 2023, numerous articles were published about the Vouga line. However, many focused on its modernization and the National Railway Plan, which were excluded from our analysis. The investments in upgrading the line highlight its importance. In 2020, the International Union of Railways named the Vouga Historic Train one of the "Dream Trains", as noted by Jornal da Bairrada. Público (January 2020) reported on the return of the historic steam train for the final Christmas season journey, while Notícias de Aveiro (February 2020) highlighted its return during Carnival and Easter. In 2021, the number of articles on the Vouga Historic Train's return increased, featuring in various media such as Time Out Lisboa, Volta ao Mundo, Público, Diário de Coimbra, SIC Notícias, Braga TV, A Notícia, NIT, Águeda TV, Diário de Aveiro, Notícias de Aveiro, Evasões, MAGG, and Aveiro Mag, among others. The language used in these articles reflected a long-awaited return, evoking a sense of travelling back in time, with some describing it as a "unique" experience. However, due to COVID-19, the Christmas edition was cancelled, as reported by Público, which also announced its return for the special Carnival edition. In 2022, more media outlets focused on tourism promotion, such as Turisver, Lisbonne Affinités, Traventia, EchoBoomer, and Sapo Viagens, writing about the return of the Vouga Historic Train and its special editions, indicating increased interest in tourism promotion. In 2023, this interest from tourism-related media solidified with continued coverage by general and local media. Notably, new sources such as Aldeias de Portugal and Centro Portugal Film Commission began highlighting this special service, suggesting growing community impact.

According to CP, in 2019, there were 15 trips with a 76% occupancy rate. In 2020, only the Carnival edition was held. In 2021, the Christmas edition ran for two days, while in 2022, the Carnival edition ran for two days and the Christmas edition for three days. Some trips were cancelled due to weather conditions, strikes, and steam locomotive breakdowns, transporting approximately 3050 passengers according to CP (Table 1).

Table 1. Editions and number of passengers of the Vouga Historic Train.

Year/Edition	Approximate Number of Passengers
2020—Carnival (February)	150
2021—May	150 + 150 + 150 + 150
2021—Christmas (December)	150 + 150
2022—Christmas (December)	150 + 150 + 150
2022—Carnival (February)	150
2023—Christmas (November, December and January)	200 × 7
Approximate total number of passengers over the 3 years	3050 [1]

[1] Approximate value. Source: [11].

On the other hand, it should be noted that the price in 2020 was 30 euros per adult and 16.50 for children aged 4 to 12; in 2023, the prices were 37 euros for adults and 22 for children, which could evidence a greater willingness to pay due to the increased demand. However, there tends to be a perception that passengers are willing to pay more when the trip is carried out with a steam locomotive than when the service is carried out by a diesel locomotive.

Considering the three pillars of sustainability: environmental, economic, and social, the results obtained from the news review and data provided by CP on the editions, approximate number of passengers, and ticket prices reveal that environmental issues are rarely mentioned despite the use of coal and diesel, likely due to the historical aspect. Regarding economic sustainability, two relevant aspects are the price increase and existing demand. Socially, the media frequently uses the name "Vouginha", reflecting the local population's preference. Some news articles highlight the community and a local association's advocacy for the service's continuity. Editions often coincide with other events in Águeda, with visitors to the museum being welcomed by musical groups and local products available for sale. It is noteworthy that from 2020 to 2023, media attention to this service has increased, indicating greater involvement.

It is also noted that the Municipality of Águeda has been committed to maintaining these special editions of the tourist train "Comboio Histórico do Vouga", as highlighted in several media outlets, through the partnership with CP, whether through the visit to the Macinhata do Vouga Museum Center, the tasting of regional sweets, and the stop at Águeda station to visit the city.

4. Discussion

Through the analysis of news articles and data provided by CP, the authors aimed to address the study's objective. The use of the Vouga Historic Train aligns with the principles of sustainability in the tourism demand of the Águeda region, primarily through social sustainability. This allows the community to preserve their railway heritage, supporting findings by Hoekstra [11,12] in the Netherlands and Niedzielski and Malecki [13,14], who suggest that railway tourism does not necessarily require environmental and economic sustainability but rather social sustainability to promote greater social cohesion. Additionally, the use of railway heritage for tourism services, such as the Vouga Historic Train, fosters community involvement through activities like singing, selling local products, and volunteering, helping to promote local culture. This perspective aligns with studies indicating that such tourism can enhance local economic development [15,16]. The demand for this service and the evolution of ticket prices highlights that investment in railway tourism stimulates the tourism market, reflected in the increased media attention from tourism-related outlets. This growing interest may sensitize public authorities to invest in railway heritage and plan tourism activities, as advocated by White [17].

5. Conclusions

The use of railway heritage in the Vouga Historic Train tourism service contributes to the sustainability of Águeda's tourism offerings by fostering local community involvement. Regular editions could help regenerate local communities and generate positive economic impacts by boosting local product sales and new business development. Tourists seek unique, past-related experiences, underscoring the importance of media and entities in promoting this service.

This study focused on a non-exhaustive news analysis to highlight key developments and media coverage. The main limitation was the inability to detail all articles. Future research could explore promotion strategies for consistent demand and tourism development. On the other hand, through the analysis carried out, it was not possible to define a consumer profile, since they do not focus on this type of content, so we consider that this could be a line of future research.

Author Contributions: Conceptualization, E.D. and D.M.P.; methodology, E.D.; software, E.D.; validation, E.D., D.M.P. and H.P.; formal analysis, H.P.; investigation, E.D.; resources, E.D.; data curation, E.D.; writing—original draft preparation, E.D.; writing—review and editing, D.M.P.; visualization, E.D.; supervision, E.D.; project administration, E.D. All authors have read and agreed to the published version of the manuscript.

Funding: This research received no external funding.

Institutional Review Board Statement: Not applicable.

Informed Consent Statement: Informed consent was obtained from all subjects involved in this study.

Data Availability Statement: Data sharing is not applicable to this abstract.

Conflicts of Interest: The authors declare no conflicts of interest.

References

1. Diário da Assembleia da República—II Série, N.º 45. 28 de Março de 1979. Available online: https://debates.parlamento.pt/catalogo/r3/dar/s2/01/03/045/1979-03-28/973?pgs=950-951&org=PLC&plcdf=true (accessed on 20 June 2024).
2. Quintas, A.; Lourencetti, F.D.L.; Cardoso de Matos, A. *Linha do Vouga—Património Imaterial*; Associação de Municípios das Terras de Santa Maria: São João da Madeira, Portugal, 2023; ISBN 978-989-53945-5-5.
3. Gazeta dos Caminhos de Ferro, n. 1363, ano LVI, 1 de Outubro de 1944, 391. Available online: https://hemerotecadigital.cm-lisboa.pt/OBRAS/GazetaCF/1944/N1363/N1363_item1/P137.html (accessed on 20 June 2024).
4. Castro, C.; Duque, A.S.; Pato, M.L. Cidades Cocriativas: O Caso de Águeda em Portugal. In *Potencialidades e Desafios do Turismo para o Desenvolvimento das Cidades 2*; de Azevedo Sampaio, E.A., Ed.; Atena Editora: Ponta Grossa, Brazil, 2022; pp. 1–13.
5. Amato, C.; Bevilacqua, G.; Ravagnan, C. The abandoned railway heritage: From problem to opportunity for the regeneration of minor historic centres. *Smart Innov. Syst. Technol.* **2021**, *178*, 1735–1745.
6. Ruhanen, L.; Moyle, C.L.; Moyle, B. New directions in sustainable tourism research. *Tour. Rev.* **2019**, *74*, 138–149. [CrossRef]
7. Streimikiene, D.; Svagzdiene, B.; Jasinskas, E.; Simanavicius, A. Sustainable tourism development and competitiveness: The systematic literature review. *Sustain. Dev.* **2021**, *29*, 259–271. [CrossRef]
8. Liu, Z. Sustainable tourism development: A critique. *J. Sustain. Tour.* **2003**, *11*, 459–475. [CrossRef]
9. UIC (International Union of Railways). Guidelines on How to Increase Attractiveness for Rail Tourism 2020. Available online: https://uic.org/passenger/rail-tourism-toprail/ (accessed on 20 June 2024).
10. Akbulut, G.; Artvinli, E. Effects of Turkish railway museums on cultural tourism. *Proc. Soc. Behav. Sci.* **2011**, *19*, 131–138.
11. Público (4 de junho de 2021). Vouguinha Volta a Fazer Aveiro—Macinhata do Vouga: É o Regresso do Comboio Histórico. Available online: https://www.publico.pt/2021/06/04/fugas/noticia/vouguinha-volta-aveiro-macinhata-vouga-regresso-comboio-historico-1965277 (accessed on 14 June 2024).
12. Hoekstra, M.S. Iconic architecture and middle class politics of memory in a deindustrialized city. *Sociology* **2020**, *54*, 693–710.
13. Niedzielski, M.A.; Malecki, E.J. Making Tracks: Rail networks in world cities. *Ann. Assoc. Am. Geogr.* **2012**, *102*, 1409–1431.
14. Hörz, P.; Richter, M. Preserved as technical monuments, run as tourist attractions: Narrow-gauge railways in the German democratic republic. *J. Transp. Hist.* **2011**, *32*, 192–213. [CrossRef]
15. Chaplin, I. Revitalizing community values through railway regeneration in the Asia Pacific region: A tourism research and education approach. In *Railway Heritage and Tourism*; Channel View Publications: Bristol, UK, 2014; pp. 115–134.
16. Jiang, P.; Shao, L.; Baas, C. Interpretation of value advantage and sustainable tourism development for railway heritage in China Based on the analytic hierarchy process. *Sustainability* **2019**, *11*, 6492. [CrossRef]
17. Luo, L.; Duan, Q.; Wang, L.; Zhao, W.; Zhuang, Y. Increased human pressures on the alpine ecosystem along the Qinghai-Tibet Railway. *Reg. Environ. Change* **2020**, *20*, 33. [CrossRef]

MDPI AG
Grosspeteranlage 5
4052 Basel
Switzerland
Tel.: +41 61 683 77 34

Proceedings Editorial Office
E-mail: proceedings@mdpi.com
www.mdpi.com/journal/proceedings

www.ingramcontent.com/pod-product-compliance
Lightning Source LLC
Chambersburg PA
CBHW062104220926
51850CB00010B/2891
* 9 7 8 3 7 2 5 8 6 0 9 1 3 *